Arne Schneck

Bounds for Optimization of the Reflection Coefficient by Constrained Optimization in Hardy Spaces

Bounds for Optimization of the Reflection Coefficient by Constrained Optimization in Hardy Spaces

by
Arne Schneck

universitätsverlag karlsruhe

Dissertation, Universität Karlsruhe (TH)
Fakultät für Mathematik
Tag der mündlichen Prüfung: 13.05.2009
Referenten: Prof. Dr. Andreas Rieder, Prof. Dr. Andreas Kirsch

Impressum

Universitätsverlag Karlsruhe
c/o Universitätsbibliothek
Straße am Forum 2
D-76131 Karlsruhe
www.uvka.de

Universitätsverlag Karlsruhe 2009
Print on Demand

ISBN: 978-3-86644-382-2

Bounds for Optimization of the Reflection Coefficient by Constrained Optimization in Hardy Spaces

Zur Erlangung des akademischen Grades eines

DOKTORS DER NATURWISSENSCHAFTEN

von der Fakultät für Mathematik der
Universität Karlsruhe (TH)
genehmigte

DISSERTATION

von
Dipl.-Math. Arne Schneck
aus Rotenburg (Wümme)

Tag der mündlichen Prüfung: 13. Mai 2009
Referent: Prof. Dr. Andreas Rieder
Korreferent: Prof. Dr. Andreas Kirsch

Contents

Introduction

> The White Rabbit put on his spectacles. "Where shall I begin, please your Majesty?" he asked.
> "Begin at the beginning," the King said, very gravely, "and go on till you come to the end: then stop."
>
> LEWIS CARROLL, *Alice's Adventures in Wonderland*

0.1 Motivation

The starting point of this thesis is *dispersion compensation for ultra-short laser pulses.* Apart from continuous wave lasers that generate a beam of light with nearly constant intensity and a very narrow bandwidth, there are lasers that emit a train of short light pulses. Such light pulses have a broad spectrum and concentrate large amounts of energy in very small time intervals. For optical frequencies, the shortest achieved pulses are well below 10 femtoseconds ($1\,\mathrm{fs} = 10^{-15}\,\mathrm{s}$), corresponding to just a few optical cycles. At a wavelength of 750 nm (near infrared), one optical cycle lasts 2.5 fs. Prominent applications of ultra-fast lasers are medical imaging [59], the observation of molecular processes on a femtosecond timescale [67] and the modification of materials with nanometer precision [35]. For an overview of the physics of ultra-fast lasers see, e.g., [50].

One of the main problems for the generation of pulses in the femtosecond regime is *dispersion*: In the air and the various optical components of a laser the different spectral components of a light pulse travel at different speeds. This leads to an unwanted spreading of the pulse. The problem gets worse for shorter pulses because shorter pulses have a broader spectrum. Dispersion thus needs to be compensated for.

A very precise control of dispersion is possible with a special kind of mirrors. These mirrors consist of a stack of thin layers of typically two dielectric (nonconducting) materials with different refractive indices, which

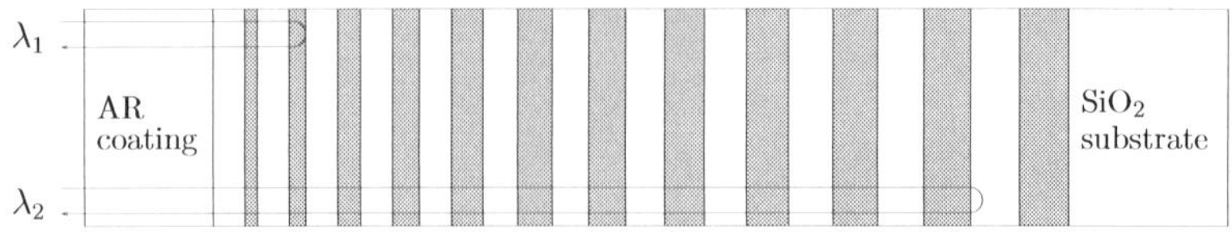

Figure 0.1: Structure of a dispersion compensating mirror (DCM). The DCM consists of a substrate and alternating layers of two dielectric materials, for example SiO_2 and TiO_2. The figure shows the structure of a *chirped mirror*: When the wavelength of an incoming wave is about twice the optical length of two adjacent layers, the wave is strongly reflected. Thus, short wavelengths (λ_1) are reflected near the top of the layer stack, while long wavelengths (λ_2) penetrate deeper into the mirror. The antireflection coating suppresses interfering reflections from the mirror surface. It also consists of alternating layers of SiO_2 and TiO_2, but the internal structure is not shown. The physical thickness of the complete layer stack is a few micrometers.

is deposited on some substrate, for example a plate of glass or SiO_2, see Figure 0.1. The stack of thin layers is called an *optical interference coating.* Typical layer materials are SiO_2 and TiO_2 with refractive indices $n_{SiO_2} \approx 1.46$ and $n_{TiO_2} \approx 2.4$ for optical frequencies. When a light wave reaches the mirror, at each layer boundary part of the wave is reflected and part of the wave is transmitted. The challenging task is now to choose the number of layers and their thicknesses in such a way that:

(a) The many reflections inside the mirror add up to give constructive interference so that the structure as a whole is highly reflective for a broad range of frequencies.

(b) Frequencies that need to be delayed penetrate deeper into the mirror than other frequencies before they are reflected.

Let us make the mirror design problem a little more precise. From a mathematical point of view, a dielectric mirror constitutes a *causal linear time-invariant (LTI) system* mapping an incoming plane wave $u_{\text{in}}(x) = e^{ikx}$ with wavenumber k to the complex conjugate* of a reflected wave $\overline{u_{\text{ref}}(x)} = \overline{R(k)}e^{ikx}$. The (complex-valued) frequency response of this LTI system is $\overline{R}$, and its complex conjugate R is called *reflection coefficient.* The modu-

*The reflected wave is $u_{\text{ref}}(x) = R(k)e^{-ikx}$. LTI systems map plane waves e^{ikx} to multiples of plane waves, $H(k)e^{ikx}$. The mapping $u_{\text{in}} \mapsto u_{\text{ref}}$ is therefore not an LTI system, but the mapping $u_{\text{in}} \mapsto \overline{u_{\text{ref}}}$ is. Details are in Chapter 2.

lus $|R(k)|$ describes the amplitude of the reflected wave, and the argument $\arg R(k)$ describes the phase shift.

For the mirror design problem one then specifies a frequency interval or a collection of frequency intervals, which we call I, and a desired complex-valued reflection coefficient R_{desired} on I. The specification is usually not given with respect to the wavenumber k, but with respect to the angular frequency $\omega = c_0 k$. Here, $c_0 = 299792458\ \text{ms}^{-1}$ is the speed of light in vacuum. Depending of the purpose of the mirror, certain deviations from the desired reflection coefficient are allowed. For example, if the mirror is used for dispersion compensation inside the laser cavity, a reflectivity $|R(\omega)| \geq 0.999$ may be required over a broad frequency interval which contains the spectrum of the generated pulse. On the other hand, if the mirror is used for external pulse compression, higher losses may be tolerable. Moreover, in some cases it may be desired to have a small frequency interval with high transparency (e.g., $|R(\omega)| \leq 0.05$), which can be used for optical pumping. The desired phase shift is usually given as a polynomial of the form

$$\arg R(\omega) = \sum_{\nu=0}^{k} \frac{1}{\nu!} D_\nu (\omega - \omega_0)^\nu.$$

with center frequency ω_0 and dispersion coefficients D_ν. Since the shape of a pulse is invariant under a constant or linear phase shift (see Chapter 1), there is some freedom in the choice of D_0 and D_1. In fact, for applications one usually only specifies the *group delay dispersion* $GDD(\omega) = \frac{d^2}{d\omega^2} \arg R(\omega)$ and requires that the GDD of the mirror does not oscillate too much around the desired GDD.

There has been an enormous amount of research on the design of mirrors with a desired reflection coefficient, for an overview see for example [24, 38, 60]. All design methods involve at some point the minimization of a *merit function* which measures how well the reflection coefficient of the mirror meets the design goal. A commonly used form of the merit function is as follows [40]. Assume that the structure of a mirror is given by a function n which describes the refractive index at different positions inside the mirror. If the mirror is made of a stack of l layers of alternating refractive indices, the space of feasible n can for example be parametrized by vectors $d = (d_1, d_2, \ldots, d_l)$ containing the layer thicknesses. Denote by R_n the reflection coefficient of a mirror with the structure given by n. A general form of a

merit function is then

$$\begin{aligned} F(n) = \sum_{j=1}^{N} w_R(\omega_j)\big||R_n(\omega_j)| - |R_{\text{desired}}(\omega_j)|\big|^{p_R} \\ + \sum_{j=1}^{N} w_{GD}(\omega_j)\big|GD_n(\omega_j) - GD_{\text{desired}}(\omega_j)\big|^{p_{GD}} \\ + \sum_{j=1}^{N} w_{GDD}(\omega_j)\big|GDD_n(\omega_j) - GDD_{\text{desired}}(\omega_j)\big|^{p_{GDD}}. \end{aligned} \tag{0.1}$$

Here, $GD_n(\omega) = \frac{\mathrm{d}}{\mathrm{d}\omega} \arg R_n(\omega)$ is the *group delay*, the meaning of which is explained in Chapter 1. Moreover, $\omega_1, \ldots, \omega_N$ are points in the frequency range of interest I, w_R, w_{GD} and w_{GDD} are weight functions, and p_R, p_{GD} and p_{GDD} are positive constants. Each of the three sums therefore has the form of a discrete weighted p-norm taken to the p-th power. Other forms of merit functions have been used, involving for example different angles of incidence or sensitivity with respect to manufacturing inaccuracies [48], but we shall not be concerned with this.

As a side note, the basic idea of optical interference coatings goes back to the beginning of the 19th century [38]. The first useful manufacturing processes were developed in the 1930s, and since then, interference coatings have been used as optical filters (e.g., edge, bandpass, beam splitters) or to reduce reflection. However, for these optical elements, the phase shift is usually not important, so in the merit function one can set $w_{GD} = w_{GDD} = 0$. Phase properties only became relevant with the introduction of dispersion compensating mirrors in 1993 [52, 53].

The problem is now that numerical minimization of the merit function F has turned out to be extremely hard. Typical dispersion compensating mirrors consist of around 40 to 100 layers of alternating material, resulting in an optimization problem in $\mathbb{R}^{40}$ to $\mathbb{R}^{100}$. This may seem small, but the merit function F is nonconvex and highly nonlinear (both as a function of the refractive profile n and the layer thicknesses d). Classical local optimization schemes such as descent methods or the Nelder-Mead simplex method quickly get stuck in local minima and are virtually useless without a decent starting design.

There are various methods to find starting designs, some of which can for example be found in the already cited books [24, 38, 60]. Unfortunately, none

of the available methods is completely general, and not for every design goal there is a method that works reliably. For example, for dispersion compensating mirrors, there is exists theory which results in so-called *double-chirped mirrors* [40], the structure of which is shown in Figure 0.1. However, the available theory does not yield any information on how to design a dispersion compensating mirror with an additional pump window. In this case, there is no way around using optimization methods.

In order to avoid getting stuck in local minima, global search methods such as simulated annealing [15] and genetic algorithms [39] have been proposed. However, they have the disadvantage of being painstakingly slow are not guaranteed to converge to a good solution [27]. The most effective general optimization procedures (i.e., which can be used without a starting design) are probably the ones based on a special method called the *needle optimization technique* [24]. The needle optimization technique is the basis of the commercial software OptiLayer [61], which has yielded some impressive results [48, 62], but also becomes slow when the number of layers is large.

To make matters worse, the efficiency of the optimization methods even depends on the proper choice of the merit function. For example, in a local minimum of the merit function, the phase properties of the reflection coefficient may be close to the design goal, but the reflectivity may be too small for practical use. In this case, one has to tweak the weight functions and start another optimization run.

0.2 Goals

The aim of this thesis is *not* to provide yet another mirror design method which might only work well in some special cases. Instead, we deal with a more basic question:

What accuracy can be obtained in principle in the mirror design problem?

The benefit of an answer to this question is obvious. At the moment, it is not quite clear when one is close to the global minimum of the merit function and should stop optimization runs. Usually, a lot of trial and error is involved during the design process, and even if one has obtained a decent solution, one cannot be sure that there is not a better solution.

In this thesis we develop a method to provide a rigorous bound on the minimum of a certain merit function, which can give an indication that one is close to an optimal solution. The properties of the reflection coefficient that we use are actually quite general: The complex conjugate of the reflection coefficient is the frequency response of a causal LTI system with no gain of energy. Our method can therefore be applied to the frequency response of any such causal LTI system with no (or limited) gain of energy.

0.3 Overview

The organization of this thesis is as follows. The first two chapters serve as an introduction and make this thesis more self-contained. In Chapter 1 we provide some physical background. We derive the *Helmholtz equation*

$$u''(x) + k^2 n^2(x) u(x) = 0 \tag{0.2}$$

from Maxwell's equations as a model for the propagation of light in layered media such as interference coatings. Moreover, we illustrate the behavior of pulses in dispersive media and explain terms like group delay (GD) and group delay dispersion (GDD). In Chapter 2 we provide some mathematical background. We introduce the *Hardy spaces* $H^p(\mathbb{D})$ on the complex unit disk and $H^p(\mathbb{C}^+)$ on the complex upper half-plane. Hardy spaces are spaces of analytic functions with restrictions on certain L^p-norms. Functions in $H^p(\mathbb{D})$ and $H^p(\mathbb{C}^+)$ have boundary values on the unit circle $\partial\mathbb{D}$ and the real line $\mathbb{R}$, respectively, and therefore the spaces $H^p(\mathbb{D})$ and $H^p(\mathbb{C}^+)$ can be identified with subspaces of $L^p(\partial\mathbb{D})$ and $L^p(\mathbb{R})$, respectively. To us, the fundamental importance of Hardy spaces lies in the fact that they are the right function spaces for frequency responses of causal LTI systems.

In Chapter 3 we rigorously define the *reflection coefficient from the right* and *from the left*, R_1 and R_2, and the *transmission coefficient* T for the Helmholtz equation and derive some of their properties. Since we are usually only interested in the reflective properties from one side, we often simply write R instead of R_2. The properties of the reflection coefficient that are most important to us in the following chapters are:

(a) Causality: The reflection coefficient lies in the Hardy space $H^\infty(\mathbb{C}^+)$.

(b) Symmetry: $\overline{R(k)} = R(-k)$ for $k \in \mathbb{R}$.

(c) No gain of energy: $|R(k)|^2 + |T(k)|^2 = 1$ for $k \in \mathbb{R}$. Especially, $|R(k)| \leq 1$ for $k \in \mathbb{R}$.

Properties (b) and (c) are not new and can for example be found in [28], but although property (a) is not unexpected, we are not aware of any proof in the literature.

Instead of using such a complicated merit function as in (0.1), we decided to use the mathematically more accessible L^p-distance, i.e., we consider the optimization problem for the reflection coefficient

$$\begin{array}{ll} \text{minimize} & \|R_n - R_{\text{desired}}\|_{L^p(I)} \\ \text{subject to} & n \in L^\infty_{a,b}(0,d), \end{array} \qquad \text{(R-OPT}_p\text{)}$$

where $1 \leq p \leq \infty$. Here, I is the frequency interval of interest, R_n is the reflection coefficient corresponding to n, $0 < a < b$, $d > 0$ and

$$L^\infty_{a,b}(0,d) = \left\{ f \in L^\infty(\mathbb{R}) : f|_{\mathbb{R}\setminus[0,d]} = 1,\ a \leq f(x) \leq b \text{ for a.a. } x \in [0,d] \right\}.$$

When a, b and d are chosen correctly, then refractive profiles $n \in L^\infty_{a,b}(0,d)$ are (in principle) physically realizable. We prove in Chapter 3 that (R-OPT$_p$) has a solution, but as we mentioned, in the general case there is no hope of actually finding a minimizing n or at least finding the minimum of (R-OPT$_p$).

Instead, in Chapters 4 and 5 we derive a *bound* of the minimum of (R-OPT$_p$). The idea is to do this by replacing the search space $\{R_n : n \in L^\infty_{a,b}(0,d)\}$ by a larger (but not too large) search space with nicer properties. Taking into account properties (a)–(c) of the reflection coefficient, it seems reasonable to consider the problem

$$\begin{array}{ll} \text{minimize} & \|R - R_{\text{desired}}\|_{L^p(I)} \\ \text{subject to} & R \in H^\infty(\mathbb{C}^+),\ |R| \leq 1,\ R \text{ real symmetric}. \end{array} \qquad (0.3)$$

However, instead of dealing with (0.3) directly, we consider the optimization problem

$$\begin{array}{ll} \text{minimize} & \|f - \varphi\|_{L^p(K)} \\ \text{subject to} & f \in E, \\ & |f| \leq g \quad \text{on } \partial\mathbb{D}. \end{array} \qquad \text{(OPT}_p\text{)}$$

Here, E is either $H^\infty(\mathbb{D})$ or $\mathcal{A}(\mathbb{D}) = H^\infty(\mathbb{D}) \cap C(\partial\mathbb{D})$. In the first case we denote the problem by (H-OPT$_p$), and in the second case we denote it by ($\mathcal{A}$-OPT$_p$). Further, $K \subset \partial\mathbb{D}$ is closed with positive measure, $g \in C(\partial\mathbb{D})$

with $g > 0$, and $\varphi \in C(K)$ such that $|\varphi| \leq g$ on K. The reason for considering (H-OPT$_p$) instead of (0.3) is that the Hardy spaces $H^p(\mathbb{D})$ and $H^p(\mathbb{C}^+)$ are isometric, but for computations it is more convenient to work in $H^p(\mathbb{D})$. Also, using the constraint $|\varphi| \leq g$ instead of $|\varphi| \leq 1$ allows more flexible modelling.

In Chapter 4 we study theoretical properties of (OPT$_p$). We prove existence and uniqueness for (H-OPT$_p$), and we show that the solution of (H-OPT$_p$) satisfies a remarkable extremal property. Moreover, we show that the infimum of ($\mathcal{A}$-OPT$_p$) is equal to the minimum of (H-OPT$_p$). This is important, because in our numerical computations we only work with smooth functions. In Chapter 5 we solve (H-OPT$_p$) numerically. We first devise a general discretization scheme and show convergence of minimum and minimizer of the discrete problem to minimum and minimizer of (H-OPT$_p$). Next, we show how to cast the discretized problem into a form that can be solved efficiently with modern numerical methods. We finish with some numerical examples. Especially, we demonstrate that our results can yield practically relevant information: We consider an example where even after long optimization runs physicists have not been able to find a refractive profile that meets a certain design goal. We will see that the "virtual" reflection coefficient that is obtained via solution of (OPT$_p$) just barely satisfies the requirements. However, because our search space is larger than the space of realizable reflection coefficients, this is a strong sign that it is not possible at all to find a refractive profile with the desired properties.

0.4 Acknowledgements

The work on this thesis was supported by a grant from the German Research Foundation (DFG) in the Research Training Group 1294 "Analysis, Simulation and Design of Nanotechnological Processes" at the Department of Mathematics, Universität Karlsruhe (TH). The financial support is gratefully acknowledged.

Especially, I would like to thank my advisor Prof. Dr. Andreas Rieder for the supervision of this thesis and for the helpful discussions that led to the improvement of some of the results. Moreover, I want to thank Prof. Dr. Andreas Kirsch for kindly agreeing to be the co-examiner of this thesis.

Further, I am obliged to Prof. Dr. Uwe Morgner from the Ultrafast Laser Optics group at the Institute for Quantum Optics, University of Hannover, for answering my questions concerning the underlying physics of this thesis and for providing me with data for numerical experiments.

Furthermore, I want to thank my present and former colleagues from the Research Training Group for valuable discussions of both mathematical and personal nature. Especially, I want to mention Alexander Bulovyatov, Christian Engström, Thomas Gauss, Armin Lechleiter and Kai Sandfort.

Finally, I thank Wolfgang Müller for giving me access to one of the nodes of the cluster at the Institute for Applied and Numerical Mathematics, which made some of the computations in Chapter 5 possible.

Chapter 1

Physical Background

> Those who like their mathematics self-contained and are not inspired by its relation to the physical world may proceed at once to the next chapter.
>
> NICHOLAS YOUNG, *An Introduction to Hilbert Space*

In this chapter we provide some physical background for this thesis. First, we derive the (one-dimensional) Helmholtz equation as a model for the propagation of electromagnetic waves in layered media. Moreover, we introduce the notion of a pulse. When a pulse travels through a linear and dispersive medium, its different spectral components travel at different speeds. This leads to a spreading of the pulse. We illustrate this effect using the example of a Gaussian pulse.

1.1 The Helmholtz equation

The evolution of electromagnetic fields in matter is governed by a set of complicated partial differential equations called Maxwell's equations [9],

$$\begin{aligned}
\nabla \times \boldsymbol{\mathcal{E}} &= -\frac{\partial}{\partial t}\boldsymbol{\mathcal{B}}, \\
\nabla \times \boldsymbol{\mathcal{H}} &= \boldsymbol{\mathcal{J}} + \frac{\partial}{\partial t}\boldsymbol{\mathcal{D}}, \\
\nabla \cdot \boldsymbol{\mathcal{D}} &= \varrho, \\
\nabla \cdot \boldsymbol{\mathcal{B}} &= 0.
\end{aligned}$$

$\boldsymbol{\mathcal{E}}$ and $\boldsymbol{\mathcal{H}}$ are called the *electric field* and the *magnetic field*, respectively. $\boldsymbol{\mathcal{D}}$ is the *electric displacement*, and $\boldsymbol{\mathcal{B}}$ is the *magnetic induction*. Furthermore, $\boldsymbol{\mathcal{J}}$ is the *electric current density*, and ρ is the *electric charge density*. All quantities depend on position $\mathbf{x} = (x, y, z) \in \mathbb{R}^3$ and time $t \in \mathbb{R}$. Quantities that are typeset in bold letters are vector-valued ($\mathbb{R}^3$), otherwise they are scalar-valued.

Additionally, some of the fields are related via *constitutive laws* (or *material equations*), which describe the reaction of the matter that occupies the domain where the electromagnetic fields live. These relations are more easily described in the frequency domain. We therefore assume that the fields are time-harmonic, i.e., that they have the form

$$\begin{aligned} \boldsymbol{\mathcal{E}}(\mathbf{x}, t) &= \mathrm{Re}(\mathrm{e}^{-i\omega t}\mathbf{E}(\mathbf{x})), & \boldsymbol{\mathcal{H}}(\mathbf{x}, t) &= \mathrm{Re}(\mathrm{e}^{-i\omega t}\mathbf{H}(\mathbf{x})), \\ \boldsymbol{\mathcal{D}}(\mathbf{x}, t) &= \mathrm{Re}(\mathrm{e}^{-i\omega t}\mathbf{D}(\mathbf{x})), & \boldsymbol{\mathcal{B}}(\mathbf{x}, t) &= \mathrm{Re}(\mathrm{e}^{-i\omega t}\mathbf{B}(\mathbf{x})), \\ \boldsymbol{\mathcal{J}}(\mathbf{x}, t) &= \mathrm{Re}(\mathrm{e}^{-i\omega t}\mathbf{J}(\mathbf{x})), & \varrho(\mathbf{x}, t) &= \mathrm{Re}(\mathrm{e}^{-i\omega t}\rho(\mathbf{x})) \end{aligned} \tag{1.1}$$

for some fixed real ω. It is also common to simply write $\boldsymbol{\mathcal{E}}(\mathbf{x}, t) = \mathrm{e}^{-i\omega t}\mathbf{E}(\mathbf{x})$ and so on. Then it is silently understood that in order to obtain the actual fields one has to take the real part. With (1.1) Maxwell's equations simplify to

$$\begin{aligned} \nabla \times \mathbf{E} &= i\omega\mathbf{B}, \\ \nabla \times \mathbf{H} &= \mathbf{J} - i\omega\mathbf{D}, \\ \nabla \cdot \mathbf{D} &= \rho, \\ \nabla \cdot \mathbf{B} &= 0. \end{aligned}$$

If the field strengths are not too large, the reaction of a material to the field is linear. Moreover, we are only going to work with isotropic materials. We then have the material equations

$$\begin{aligned} \mathbf{D} &= \epsilon\mathbf{E}, \\ \mathbf{B} &= \mu\mathbf{H}. \end{aligned}$$

The function $\epsilon = \epsilon(\mathbf{x}, \omega)$ is called *(electric) permittivity*, and $\mu = \mu(\mathbf{x}, \omega)$ is called *(magnetic) permeability*. In this thesis, we only consider nonmagnetic materials, in which case $\mu \equiv \mu_0 \approx 1.2566 \times 10^{-6}\ \mathrm{Hm}^{-1}$, the permeability of free space. Further, $\mathbf{J}$ and $\mathbf{E}$ are related via

$$\mathbf{J} = \sigma\mathbf{E},$$

where $\sigma = \sigma(\mathbf{x}, \omega)$ is the *conductivity*. We are only going to deal with nonconducting materials, so we can set $\sigma \equiv 0$. Finally, there are no external

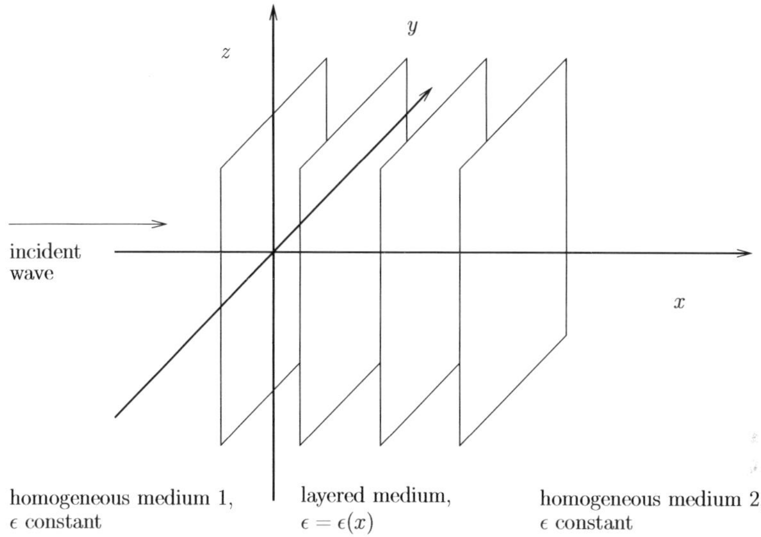

Figure 1.1: Layered medium.

charges involved in our setup, i.e., $\rho \equiv 0$. Under all these assumptions on the materials, we end up with

$$\nabla \times \mathbf{E} = i\omega\mu_0\mathbf{H}, \tag{1.2}$$

$$\nabla \times \mathbf{H} = -i\omega\epsilon\mathbf{E}, \tag{1.3}$$

$$\nabla \cdot \epsilon\mathbf{E} = 0, \tag{1.4}$$

$$\nabla \cdot \mathbf{H} = 0. \tag{1.5}$$

In the case of a layered medium like a dielectric mirror, the above equations can be simplified further. Let us assume that the medium is homogeneous in the y- and z-direction so that if ω is fixed, the permittivity ϵ depends only on x, see Figure 1.1. For simplicity, we assume that the incident electromagnetic field propagates in x-direction and is linearly polarized with the electric field pointing in y-direction and the magnetic field pointing in z-direction, i.e.,

$$\mathbf{E}(\mathbf{x}) = E(x)\mathbf{e}_y, \tag{1.6}$$

$$\mathbf{H}(\mathbf{x}) = H(x)\mathbf{e}_z. \tag{1.7}$$

One immediately checks that (1.4) and (1.5) are then automatically fulfilled. Moreover, it follows readily that $\nabla \times \mathbf{E} = E'(x)\mathbf{e}_z$ and $\nabla \times \mathbf{H} = -H'(x)\mathbf{e}_y$ such that (1.2) and (1.3) yield

$$E'(x) = i\omega\mu_0 H(x), \tag{1.8}$$
$$H'(x) = i\omega\epsilon(x)E(x). \tag{1.9}$$

Using (1.9) in (1.8), we arrive at

$$E''(x) + \omega^2\mu_0\epsilon(x)E(x) = 0. \tag{1.10}$$

We finally rearrange this equation into a more common form. Let $n(x) = \sqrt{\epsilon(x)/\epsilon_0}$ be the *refractive index* of the material at position x. Here, $\epsilon_0 \approx 8.8542 \times 10^{-12}\ \mathrm{Fm}^{-1}$ is the permittivity of free space. With $c_0 = 1/\sqrt{\mu_0\epsilon_0}$, the *speed of light in vacuum*, and $k = \omega/c_0$, (1.10) transforms into

$$E''(x) + k^2n^2(x)E(x) = 0, \tag{1.11}$$

the one-dimensional *Helmholtz equation.* If we know the refractive profile n of our layered medium and the angular frequency ω (or k) of the incident wave as well as some initial conditions, we can then solve the above equation for E and get original $\boldsymbol{\mathcal{E}}$- and $\boldsymbol{\mathcal{H}}$-fields via (1.8), (1.6) and (1.7), and (1.1).

Layered media like dielectric mirrors are not only used for perpendicularly incident waves, but also at oblique angles of incidence. In this case one has to decompose the incident fields into two components, the S-component and the P-component, see Figure 1.2. The letters S and P come from the German words *senkrecht* (perpendicular) and *parallel*. In the S-polarization case, the $\boldsymbol{\mathcal{E}}$-field has only a y-component, i.e., it is perpendicular to the plane of incidence, the xz-plane. In the P-polarization case, the $\boldsymbol{\mathcal{E}}$-field has only x- and z-components, i.e., it is parallel to the plane of incidence. For both polarizations one can reduce (1.2)–(1.5) to an equation which also has the form of the Helmholtz equation. Thus, the mathematical model is the same. For details see, e.g., [24].

1.2 Pulses

There does not seem to be a precise mathematical definition of a light pulse in the literature. A pulse is generally taken to mean a function f such that both

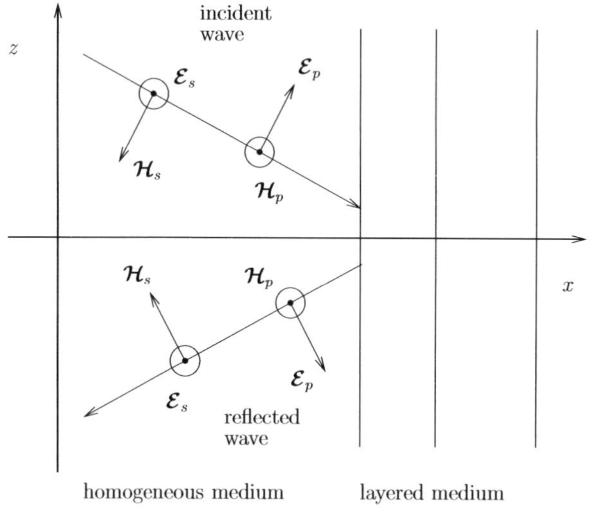

Figure 1.2: Direction of electric and magnetic field in the S-polarization case (left) and P-polarization case (right).

f and its Fourier transform $\widehat{f}$ are localized in some sense. More concretely, think of a pulse as a rapidly varying function like a plane wave $\mathrm{e}^{i\omega_0 x}$ multiplied by some slowly varying bell-shaped function m. The rapidly varying function is called the *carrier wave*, and m is called the *envelope*. If m is a Gaussian function, $m(x) = \mathrm{e}^{-bx^2}$ for some $b > 0$, then $f(x) = m(x)\mathrm{e}^{i\omega_0 x} = \mathrm{e}^{-bx^2}\mathrm{e}^{i\omega_0 x}$ is called a *Gaussian pulse*. The Fourier transform of a Gaussian pulse is a Gaussian function,

$$\widehat{f}(\omega) = \int_{\mathbb{R}} f(x)\mathrm{e}^{-i\omega x}\,\mathrm{d}x = \left(\frac{\pi}{b}\right)^{1/2} \mathrm{e}^{-\frac{1}{4b}(\omega-\omega_0)^2},$$

so $\widehat{f}$ is again localized. Notice that the shorter the Gaussian pulse is, i.e., the larger b is, the wider its Fourier transform becomes. This is the consequence of a more general principle, the *uncertainty principle* [10, 23]. For a non-zero function $f \in L^2(\mathbb{R})$ denote by

$$\mu_f = \frac{1}{\|f\|^2_{L^2(\mathbb{R})}} \int_{\mathbb{R}} x|f(x)|^2\,\mathrm{d}x$$

the *center* and by

$$\sigma_f = \left(\frac{1}{\|f\|^2_{L^2(\mathbb{R})}} \int_{\mathbb{R}} (x - \mu_f)^2 |f(x)|^2 \, \mathrm{d}x \right)^{1/2}$$

the *root mean square (RMS) width* (supposing that the integrals exist).

Theorem 1.1 (Heisenberg's inequality). *Supposing that the above integrals exist, it must hold true that*

$$\sigma_f \sigma_{\hat{f}} \geq \frac{1}{2}.$$

Equality holds true if and only if f is a shifted and scaled version of a Gaussian pulse, i.e., $f(x) = a\mathrm{e}^{-b(x-c)^2}\mathrm{e}^{i\omega_0 x}$ for some $a \in \mathbb{C} \setminus \{0\}$, $b > 0$ and c, $\omega_0 \in \mathbb{R}$.

So the shorter a pulse is in the time (or space) domain, the wider it must be in the frequency domain, and the more narrow it is in the frequency domain, the longer it must be in the time (or space) domain.

1.3 Dispersion

The speed at which a plane wave propagates in a homogeneous medium depends on the refractive index. Suppose the space is filled with some medium with refractive index n_0. The general solution of the Helmholtz equation (1.11) is then $E(x) = \alpha \mathrm{e}^{ikn_0 x} + \beta \mathrm{e}^{-ikn_0 x}$, and for the electric field we have $\boldsymbol{\mathcal{E}}(\mathbf{x}, t) = E(x)\mathrm{e}^{-i\omega t}\mathbf{e}_y$. Let us just look at

$$\psi(x,t) = E(x)\mathrm{e}^{-i\omega t} = \alpha \mathrm{e}^{i(kn_0 x - \omega t)} + \beta \mathrm{e}^{i(-kn_0 x - \omega t)}.$$

The phase of the first term is constant if $kn_0 x = \omega t$, i.e., $x = \frac{\omega}{kn_0}t = \frac{c_0}{n_0}t$. So the first term is a plane wave travelling to the right with speed c_0/n_0. Similarly, the second term is a plane wave travelling to the left with speed c_0/n_0. Since the refractive index of a material depends on the frequency ω, plane waves of different frequencies propagate at different speeds. This effect is called *dispersion*.

We illustrate the effect of dispersion using the example of a Gaussian pulse that propagates through some medium. We start out with a pulse of the form

$$f(x) = \mathrm{e}^{-bx^2}\mathrm{e}^{i\omega_0 x}. \tag{1.12}$$

The pulse can also be written as the superposition of plane waves,

$$f(x) = \int_{\mathbb{R}} \frac{1}{2\pi} \widehat{f}(\omega) \mathrm{e}^{i\omega x} \,\mathrm{d}\omega.$$

After the pulse has travelled for some time t_0, each plane wave $\mathrm{e}^{i\omega x}$ has experienced a phase shift ϕ, the size of which depends on the frequency ω. The pulse then has the form

$$f_{t_0}(x) = \int_{\mathbb{R}} \frac{1}{2\pi} \widehat{f}(\omega) \mathrm{e}^{i\omega x} \mathrm{e}^{-i\phi(\omega)} \,\mathrm{d}\omega.$$

Since $\widehat{f}$ is localized about ω_0, one usually assumes that ϕ has a Taylor expansion around ω_0,

$$\phi(\omega) = \sum_{\nu=0}^{\infty} \frac{1}{\nu!} \phi^{(\nu)}(\omega_0)(\omega - \omega_0)^j. \tag{1.13}$$

The numbers $\phi^{(\nu)}(\omega_0)$ are abbreviated by D_ν and are called *dispersion coefficients*. The broader the pulse is, the stronger it is localized about ω_0, and therefore the earlier we can truncate the above Taylor series and still get a good approximation of f_{t_0}.

Let us assume that the pulse is not too short so that a linear approximation suffices, i.e., $\phi(\omega) = D_0 + D_1(\omega - \omega_0)$ with some D_0, $D_1 \in \mathbb{R}$. A straightforward calculation then shows that

$$f_{t_0}(x) = \mathrm{e}^{i(\omega_0(x - D_1) - D_0)} \mathrm{e}^{-b(x - D_1)^2} = \mathrm{e}^{-iD_0} f(x - D_1).$$

This means that the envelope of the pulse has been shifted by D_1. Because $\phi'(\omega_0) = D_1$, the function ϕ' is called *group delay*. The term e^{-iD_0} merely shifts the carrier wave under the envelope.

If the pulse is relatively short, then $\widehat{f}$ is only weakly localized about ω_0 and one needs a higher order approximation of ϕ. Let us illustrate the effect of $D_2 = \phi''(\omega_0)$. The function ϕ'' is called *group delay dispersion (GDD)*. Because we already know the effect of D_0 and D_1, we assume $\phi(\omega) =$

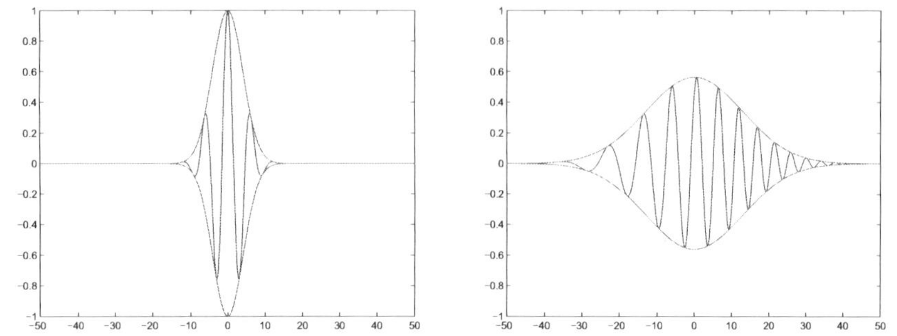

Figure 1.3: GDD causes spreading of the pulse and creates a chirp: Higher frequencies are found on the right hand side of the pulse, and lower frequencies are found on the left hand side of the pulse. The parameters were $\omega_0 = 1$, $b = 0.03$ and $D_2 = 25$.

$D_2(\omega - \omega_0)^2$. After the phase shift the pulse (1.12) has the form

$$\begin{aligned} f_{t_0}(x) &= \int_{\mathbb{R}} \frac{1}{2\pi} \widehat{f}(\omega) \mathrm{e}^{i\omega x} \mathrm{e}^{-i\phi(\omega)} \,\mathrm{d}\omega \\ &= \int_{\mathbb{R}} \frac{1}{2(b\pi)^{1/2}} \mathrm{e}^{-\frac{1}{4b}(\omega-\omega_0)^2} \mathrm{e}^{i\omega x} \mathrm{e}^{-iD^2(\omega-\omega_0)^2} \,\mathrm{d}\omega \\ &= \frac{1}{\sqrt{4D_2 bi + 1}} \exp\left(\frac{-bx^2 - 4D_2 b\omega_0 x + i\omega_0 x}{4D_2 bi + 1} \right) \\ &= \frac{1}{\sqrt{4D_2 bi + 1}} \exp\left(\frac{-bx^2}{1 + 16D_2^2 b^2} \right) \exp\left(i\omega_0 x + i\frac{4b^2 D_2}{1 + 16D_2^2 b^2} x^2 \right). \end{aligned} \tag{1.14}$$

The first exponential factor shows that a nonzero GDD causes a spreading of the pulse, see Figure 1.3. The pulse becomes twice as wide if $1 + 16D_2^2 b^2 = 4$, or $|D_2| = \frac{\sqrt{3}}{4b}$. So the shorter a pulse is (i.e., the larger b is), the more sensitive it is to GDD. Moreover, the quadratic term in the second factor shows that there is a *frequency chirp*: The *instantaneous frequency* ω_{inst} of a pulse f is defined by $\omega_{\text{inst}}(x) = \frac{\mathrm{d}}{\mathrm{d}x} \arg f(x)$. For the above pulse f_{t_0} from (1.14) we have

$$\omega_{\text{inst}}(x) = \omega_0 + \frac{8b^2 D_2}{1 + 16D_2^2 b^2} x,$$

that is, the instantaneous frequency varies linearly in x.

The shorter a pulse is, the later we may truncate the series in (1.13). The higher order derivatives of ϕ are called *third order dispersion* (for $\phi^{(3)}$), *fourth order dispersion* (for $\phi^{(4)}$), and so on. The influence of $D_\nu = \phi^{(\nu)}(\omega_0)$ is just not as easily described for $\nu \geq 3$. For the generation of pulses in the femtosecond regime, it is not uncommon that the phase shift that has to be compensated for is given as a sixth order polynomial [40], i.e., the dispersion coefficients up to D_6 are taken into account.

Chapter 2

Hardy Spaces, LTI Systems and the Paley-Wiener Theorem

> There seems to be no part of (so-called pure) mathematics that is not in immediate danger of being applied.
>
> ---
>
> MICHIEL HAZEWINKEL, preface to *Complex analytic sets* by E. M. CHIRKA

In this chapter we provide some mathematical background. In the first section we introduce the *Hardy spaces* $H^p(\mathbb{D})$ on the complex unit disk and $H^p(\mathbb{C}^+)$ on the complex upper half-plane. Hardy spaces are spaces of analytic functions with certain L^p-norm restrictions. These spaces are well-known and are used in some areas of both pure and applied mathematics. Nevertheless, in order to make this thesis more self-contained and for later reference, we state the definitions of Hardy spaces and some of their very basic properties. To the interested reader we warmly recommend the excellent introductory book by HOFFMAN [32]. An even more basic introduction to only $H^p(\mathbb{D})$ (but with applications in the spirit of Chapters 4 and 5 of this thesis) can be found in the book by YOUNG [66]. A more complete account of Hardy spaces is for example given by GARNETT [25].

Hardy spaces are so fundamental for this thesis because they occur in the context of causal linear time-invariant (LTI) systems, which we briefly introduce in the second section of this chapter. Causal LTI systems are characterized by their frequency response H, which has the property that $\operatorname{supp} \widehat{H^\circ} \subset [0, \infty)$, where $H^\circ(\omega) = H(-\omega)$. The relationship to Hardy spaces is due to the Paley-Wiener Theorem, which gives (in its classical version) the identification

$$H^2(\mathbb{C}^+) = \{f \in L^2(\mathbb{R}) : \operatorname{supp} \widehat{f} \subset [0, \infty)\}.$$

Such an identification still holds true for $H^p(\mathbb{C}^+)$ with $p \neq 2$, which is not surprising, but we do not know of any reference where this is explicitly stated and proved. We fill this gap in the last section of this chapter.

2.1 Hardy spaces

There are basically two classical kinds of Hardy spaces, those on the complex unit disk $\mathbb{D} = \{z \in \mathbb{C} : |z| < 1\}$, and those on the complex upper half-plane $\mathbb{C}^+ = \{z \in \mathbb{C}^+ : \operatorname{Im} z > 0\}$. Both kinds of spaces are important for us: We will see in Chapter 3 that the reflection coefficient for the Helmholtz equation lies in the Hardy space $H^\infty(\mathbb{C}^+)$. However, for computations it is easier to work in $H^p(\mathbb{D})$. In Chapters 4 and 5 we will therefore use the latter spaces.

2.1.1 Hardy spaces on the disk: $H^p(\mathbb{D})$

Hardy spaces on the unit disk $\mathbb{D} = \{z \in \mathbb{C} : |z| < 1\}$ are defined as follows.

Definition 2.1. *Let $1 \leq p \leq \infty$. The* Hardy space $H^p(\mathbb{D})$ *is*

$$H^p(\mathbb{D}) = \left\{ f : \mathbb{D} \to \mathbb{C} : f \text{ is analytic on } \mathbb{D},\ \sup_{0 \leq r < 1} \|f_r\|_{L^p(-\pi,\pi)} < \infty \right\},$$

where $f_r(\vartheta) = f(re^{i\vartheta})$.

Although by the above definition functions from $H^p(\mathbb{D})$ a priori live only on the open unit disk $\mathbb{D}$, one can identify $H^p(\mathbb{D})$ with a subspace of $L^p(\partial\mathbb{D})$. A function from $L^p(\partial\mathbb{D})$ gives rise to a function on the unit disk as follows.

Definition 2.2. *Let $0 < r < 1$. The function*

$$P_r(\vartheta) = \frac{1 - r^2}{1 - 2r\cos\vartheta + r^2}$$

is called Poisson kernel *for the disk.*

Poisson's kernel is an approximate identity, i.e., it holds true that

(a) $P_r(\vartheta) \geq 0$.

(b) $\frac{1}{2\pi}\int_{-\pi}^{\pi} P_r(\vartheta)\,\mathrm{d}\vartheta = 1$ for $0 \leq r < 1$.

(c) If $0 < \delta < \pi$, then $\lim_{r \nearrow 1} \sup_{\vartheta \in [-\pi,\pi]\setminus[-\delta,\delta]} |P_r(\vartheta)| = 0$.

The next theorem states that by convolving a function from $L^p(\partial\mathbb{D})$ with the Poisson kernel, one obtains a function that is *harmonic* on the disk $\mathbb{D}$. In the following, it is convenient to identify $L^p(\partial\mathbb{D})$ with $L^p(-\pi,\pi)$. Especially, if $\vartheta \in [-\pi,\pi]$ and $f \in L^p(\partial\mathbb{D})$, then we also write $f(\vartheta)$ instead of $f(\mathrm{e}^{i\vartheta})$.

Theorem 2.3. *Let $1 \le p \le \infty$ and let $F \in L^p(\partial\mathbb{D})$. Define f on the disk by*

$$f(r\mathrm{e}^{i\vartheta}) = \frac{1}{2\pi}\int_{-\pi}^{\pi} F(t)P_r(\vartheta - t)\,\mathrm{d}t.$$

Then f is harmonic on $\mathbb{D}$, and we have the following behavior towards the boundary.

(a) *$f_r \to F$ a.e. as $r \nearrow 1$. Here again, $f_r(\vartheta) = f(r\mathrm{e}^{i\vartheta})$.*

(b) *$f_r \in L^p(-\pi,\pi)$ for every $0 \le r < 1$, and $\sup_{0\le r<1} \|f_r\|_{L^p(-\pi,\pi)} < \infty$. In fact, $\|f_r\|_{L^p(-\pi,\pi)}$ is a increasing function in r, that is, $r_1 < r_2$ implies $\|f_{r_1}\|_{L^p(-\pi,\pi)} \le \|f_{r_2}\|_{L^p(-\pi,\pi)}$.*

(c) *If $1 \le p < \infty$, then $f_r \to F$ in $L^p(-\pi,\pi)$ as $r \nearrow 1$. If $p = \infty$, then $f_r \overset{*}{\rightharpoonup} F$ in $L^\infty(-\pi,\pi)$ as $r \nearrow 1$. If $F \in C(\partial\mathbb{D})$, then $f_r \to F$ uniformly as $r \nearrow 1$.*

We also say that f *has boundary values* F on the circle. The function f is called the *Poisson integral* of F. Notice again carefully that the theorem only states that f is *harmonic* on the disk. In order for f to be in a Hardy space, it must be *analytic* on the disk. The following theorem states that functions from $H^p(\mathbb{D})$ have boundary values in $L^p(\partial\mathbb{D})$, and that the Poisson integral of the boundary values is the original function from $H^p(\mathbb{D})$.

Theorem 2.4. *Let $1 \le p \le \infty$ and $f \in H^p(\mathbb{D})$.*

(a) *f has boundary values on the circle, i.e., the functions $f_r(\vartheta) = f(\mathrm{e}^{i\vartheta})$ converge a.e. to some function F on the circle.*

(b) *The function F lies in $L^p(\partial\mathbb{D})$, and f is the Poisson integral of its boundary values, i.e., we have the representation*

$$f(r\mathrm{e}^{i\vartheta}) = \frac{1}{2\pi}\int_{-\pi}^{\pi} F(t)P_r(\vartheta - t)\,\mathrm{d}t.$$

(c) $H^p(\mathbb{C}^+)$ *is a Banach space with norm*

$$\|f\|_{H^p(\mathbb{D})} = \sup_{0<r<1} \|f_r\|_{L^p(-\pi,\pi)} = \|f\|_{L^p(\partial\mathbb{D})}.$$

Thus, $H^p(\mathbb{D})$ can be identified with the subspace of $L^p(\partial\mathbb{D})$ whose Poisson integrals are not only harmonic, but even analytic on the disk. Notice that we already used the identification of functions in $H^p(\mathbb{D})$ with their boundary values in the last part of the theorem, where we wrote $\|f\|_{L^p(\partial\mathbb{D})}$ instead of $\|F\|_{L^p(\partial\mathbb{D})}$.

There is a slightly more concrete characterization of $H^p(\mathbb{D})$ as a subspace of $L^p(\partial\mathbb{D})$.

Theorem 2.5. *The characterization*

$$H^p(\mathbb{D}) = \left\{ f \in L^p(\partial\mathbb{D}) : \widehat{f}_k = 0 \text{ for integers } k < 0 \right\}$$

holds true. Here, $\widehat{f}_k = \int_{-\pi}^{\pi} f(\vartheta)\mathrm{e}^{-ik\vartheta}\,\mathrm{d}\vartheta$, $k \in \mathbb{Z}$, *are the Fourier coefficients of* f.

This characterization is fairly suggestive. A function f that is analytic on the unit disk can be written as a power series $f(z) = \sum_{k=0}^{\infty} a_k z^k$. Assume that f is analytic on the closed disk. Then the series also converges on $\partial\mathbb{D}$, i.e., $f(\mathrm{e}^{i\vartheta}) = \sum_{k=0}^{\infty} a_k \mathrm{e}^{ik\vartheta}$ for $\mathrm{e}^{i\vartheta} \in \partial\mathbb{D}$. The Fourier coefficients of f as a function on $\partial\mathbb{D}$ are just $\widehat{f}_k = a_k$ for integers $k \geq 0$ and $\widehat{f}_k = 0$ for integers $k < 0$. This makes the inclusion "$\subset$" in the above theorem plausible. On the other hand, take a function $f \in L^p(\partial\mathbb{D})$ with $\widehat{f}_k = 0$ for integers $k < 0$. Then the sequence $(\widehat{f}_k)_{k\in\mathbb{Z}}$ of Fourier coefficients of f is bounded, whence $\sum_{k=0}^{\infty} \widehat{f}_k z^k$ converges uniformly on compact subsets of $\mathbb{D}$ and is therefore analytic on $\mathbb{D}$. Indeed, the function thus defined on the disk is equal to the Poisson integral of f. Of course, the whole story is more involved, and the proof of Theorem 2.5 is not easy, especially for the case $p = 1$. The interested reader will find a detailed presentation in [32, Chapters 1–4].

Functions in a Hardy space are uniquely determined by their values on a set of positive measure [25, Chapter II, Corollary 4.2].

Theorem 2.6. *If* $f \in H^p(\mathbb{D})$, $1 \leq p \leq \infty$, *and* $f = 0$ *on a set* $K \subset \partial\mathbb{D}$ *of positive (Lebesgue) measure, then* $f \equiv 0$.

Finally, by $\mathcal{A}(\mathbb{D}) = H^\infty(\mathbb{D}) \cap C(\partial\mathbb{D})$ we denote the space of functions in $H^\infty(\mathbb{D})$ with continuous boundary values. $\mathcal{A}(\mathbb{D})$ is called the *disk algebra*.

2.1.2 Hardy spaces on the half-plane: $H^p(\mathbb{C}^+)$

The second kind of Hardy spaces that we need are those on the complex upper half-plane $\mathbb{C}^+ = \{z \in \mathbb{C}^+ : \operatorname{Im} z > 0\}$.

Definition 2.7. *Let* $1 \leq p \leq \infty$. *The* Hardy space $H^p(\mathbb{C}^+)$ *is*

$$H^p(\mathbb{C}^+) = \left\{ f : \mathbb{C}^+ \to \mathbb{C} : f \text{ is analytic on } \mathbb{C}^+, \ \sup_{y>0} \|f(\cdot + iy)\|_{L^p(\mathbb{R})} < \infty \right\}.$$

The theory for $H^p(\mathbb{C}^+)$ is very similar to the theory for $H^p(\mathbb{D})$. Functions defined on $\mathbb{R}$ $(= \partial\mathbb{C}^+)$ give rise to functions on $\mathbb{C}^+$ via convolution with Poisson's kernel for the upper half-plane.

Definition 2.8. *Let* $y > 0$. *The function*

$$P_y(t) = \frac{y}{t^2 + y^2}, \qquad t \in \mathbb{R},$$

is called Poisson kernel *for the upper half-plane.*

The following two theorems are the analogon to Theorem 2.3.

Theorem 2.9. *Let* $F : \mathbb{R} \to \mathbb{C}$ *be integrable with respect to the measure* $\frac{1}{1+t^2}\,\mathrm{d}t$. *Define* f *on the complex upper half-plane by*

$$f(x + iy) = \frac{1}{\pi} \int_{\mathbb{R}} F(t) P_y(x - t)\,\mathrm{d}t, \qquad x \in \mathbb{R},\ y > 0. \tag{2.1}$$

Then f *is harmonic on* $\mathbb{C}^+$, *and* $f(\cdot + iy) \to F$ *a.e. as* $y \searrow 0$.

Just like in the case of the disk we say that f *has boundary values* F on the real line, and that f is the Poisson integral of F. Again, the theorem only states that f is *harmonic* on $\mathbb{C}^+$, not analytic. Notice that the theorem is especially true for $F \in L^p(\mathbb{R})$, $1 \leq p \leq \infty$. If $F \in L^p(\mathbb{R})$, a stronger result holds true.

Theorem 2.10. *Let* $1 \leq p \leq \infty$ *and let* $F \in L^p(\mathbb{R})$. *Define* f *on* $\mathbb{C}^+$ *as in* (2.1).

(a) *For each* $y > 0$, *the function* $f_y(x) = f(x + iy)$ *is in* $L^p(\mathbb{R})$. *The* $L^p(\mathbb{R})$-*norms of* f_y *are bounded for* $y > 0$. *In fact,* $\|f_y\|_{L^p(\mathbb{R})}$ *is a decreasing function of* y *for* $y > 0$.

(b) *If* $1 \leq p < \infty$, *then* $f_y \to F$ *in* $L^p(\mathbb{R})$ *as* $y \searrow 0$. *If* $p = \infty$, *then* $f_y \overset{*}{\rightharpoonup} F$ *in* $L^\infty(\mathbb{R})$ *as* $y \searrow 0$. *If* $F \in C(\mathbb{R})$, *then* $f_y \to F$ *uniformly as* $y \searrow 0$.

The following theorem is the analogon to Theorem 2.4. It states that functions from $H^p(\mathbb{C}^+)$ have boundary values in $L^p(\mathbb{R})$, and that functions from $H^p(\mathbb{C}^+)$ are the Poisson integral of their boundary values.

Theorem 2.11. *Let* $1 \leq p \leq \infty$ *and* $f \in H^p(\mathbb{C}^+)$.

(a) f *has boundary values on the real line, i.e., the functions* $f(\cdot + iy)$ *converge a.e. to some function* F *on the real line as* $y \searrow 0$.

(b) *The function* F *lies in* $L^p(\mathbb{R})$, *and* f *is the Poisson integral of its boundary values, i.e., we have the representation*

$$f(x + iy) = \frac{1}{\pi} \int_{\mathbb{R}} F(t) P_y(x - t)\, \mathrm{d}t, \qquad x \in \mathbb{R},\ y > 0.$$

(c) $H^p(\mathbb{C}^+)$ *is a Banach space with norm*

$$\|f\|_{H^p(\mathbb{C}^+)} = \sup_{y>0} \|f(\cdot + iy)\|_{L^p(\mathbb{R})} = \|f\|_{L^p(\mathbb{R})}.$$

Like in the case of the disk, $H^p(\mathbb{C}^+)$ is identified with the subspace of $L^p(\mathbb{R})$ whose Poisson integrals are analytic, and we already used the identification in part (c) of the theorem, where we wrote $\|f\|_{L^p(\mathbb{R})}$ instead of $\|F\|_{L^p(\mathbb{R})}$.

The relation between $H^p(\mathbb{D})$ and $H^p(\mathbb{C}^+)$ is as follows. The Möbius transformation $w \mapsto \frac{iw+1}{-iw+1}$ maps $\mathbb{C}^+$ conformally to $\mathbb{D}$. Given a function $f \in H^p(\mathbb{D})$, the function $g : w \mapsto f\left(\frac{iw+1}{-iw+1}\right)$ is analytic on $\mathbb{C}^+$. However, it may not be true that $\sup_{y>0} \|g(\cdot + iy)\|_{L^p(\mathbb{R})} < \infty$. In order for g to be in a Hardy space, an additional decay factor is needed. In fact, we have the following theorem.

Theorem 2.12. *The mapping*

$$\begin{cases} T_p : H^p(\mathbb{D}) & \longrightarrow & H^p(\mathbb{C}^+) \\ \qquad\quad g & \longmapsto & f,\ f(w) = 2^{-1/p}(i + w)^{-2/p} g\left(\frac{iw+1}{-iw+1}\right). \end{cases}$$

is an isomorphism. If the norm on $H^p(\mathbb{D})$ *is normalized to* $\|f\|^p_{H^p(\mathbb{D})} = \int_{-\pi}^{\pi} |f(\mathrm{e}^{i\vartheta})|^p\, \mathrm{d}\vartheta$ *(i.e., the integral is not taken with respect to normalized Lebesgue measure), then the mapping is even an isometry.*

For us, the importance of the theorem is due to the fact that the Hardy spaces $H^p(\mathbb{C}^+)$ occur in our applications. However, it is much easier to do computations in $H^p(\mathbb{D})$. The isometry can therefore be used to transport functions from $H^p(\mathbb{C}^+)$ to $H^p(\mathbb{D})$ and back.

There is a characterization of $H^2(\mathbb{C}^+)$ as a subspace of $L^2(\mathbb{R})$ similar to that of Theorem 2.5. We use the Fourier transform in the version

$$\widehat{f}(t) = \int_{\mathbb{R}} f(x) \mathrm{e}^{-itx} \, \mathrm{d}x,$$

so that the inverse Fourier transform is given by

$$\widetilde{f}(t) = \frac{1}{2\pi} \int_{\mathbb{R}} f(x) \mathrm{e}^{itx} \, \mathrm{d}x.$$

Theorem 2.13 (Paley-Wiener). *We have the characterization*

$$H^2(\mathbb{C}^+) = \left\{ f \in L^2(\mathbb{R}) : \operatorname{supp} \widehat{f} \subset [0, \infty) \right\}.$$

The Paley-Wiener theorem is a classical result [25, 32, 36], which has been generalized in numerous ways [7, 12, 46]. However, we are not aware of any reference where it is stated or proved in the above form for $H^p(\mathbb{C}^+)$ and $L^p(\mathbb{R})$ with general $1 \leq p \leq \infty$. We fill this gap in the last section of this chapter.

2.2 LTI systems

Causal LTI systems and the Paley-Wiener theorem are the reason why we deal with Hardy spaces. By $\mathcal{S}'(\mathbb{R})$ we denote the space of tempered distributions over $\mathbb{R}$. A formal definition of an LTI system is as follows [10].

Definition 2.14. *Let $X \subset \mathcal{S}'(\mathbb{R})$ be a translation invariant subspace, i.e., $f \in X$ implies $f(\cdot - t) \in X$ for all $t \in \mathbb{R}$. A* linear time-invariant (LTI) system *is a mapping $L : X \subset \mathcal{S}'(\mathbb{R}) \to \mathcal{S}'(\mathbb{R})$ that is linear, continuous and time-invariant, i.e., $L(f(\cdot - t)) = (Lf)(\cdot - t)$ for all $t \in \mathbb{R}$.*

The system is called causal *if for all $t_0 \in \mathbb{R}$ it holds that $f(t) = 0$ for $t < t_0$ implies $Lf(t) = 0$ for $t < t_0$.*

It is not hard to see that if $\mathrm{e}^{i\omega\cdot} \in X$, then $L(\mathrm{e}^{i\omega\cdot}) = H(\omega)\mathrm{e}^{i\omega\cdot}$ for some $H(\omega) \in \mathbb{C}$. The function H is called *frequency response*. Moreover, if $\delta_0 \in X$

(Dirac impulse), then $h = L\delta_0$ is called *impulse response*, and one can show that

$$Lf = h * f, \tag{2.2}$$

where $*$ denotes convolution. We then have $H = \widehat{h}$, and the system (2.2) is causal if and only if $\operatorname{supp} h \subset [0, \infty)$, or, equivalently, $\operatorname{supp} \widehat{H^\circ} \subset [0, \infty)$, where $H^\circ(t) = H(-t)$.

Examples of frequency responses of causal LTI systems that occur in the context of electromagnetics are the following.

Example 2.15 (Permittivity). *Assume that the material that occupies some space is linear and isotropic. We stated in Chapter 1 that if the $\boldsymbol{\mathcal{E}}$-field is time-harmonic, i.e., if it has the form*

$$\boldsymbol{\mathcal{E}}(\boldsymbol{x}, t) = \mathrm{e}^{-i\omega t}\boldsymbol{E}(\boldsymbol{x}),$$

then the $\boldsymbol{\mathcal{D}}$-field satisfies

$$\boldsymbol{\mathcal{D}}(\boldsymbol{x}, t) = \mathrm{e}^{-i\omega t}\epsilon(\boldsymbol{x}, \omega)\boldsymbol{E}(\boldsymbol{x}).$$

Thus, if we fix some $\boldsymbol{x} \in \mathbb{R}^3$ and write $\boldsymbol{\mathcal{E}}(\boldsymbol{x}, t) = E(t)\boldsymbol{u}$ for some $\boldsymbol{u} \in \mathbb{R}^3$ and $\boldsymbol{\mathcal{D}}(\boldsymbol{x}, t) = D(t)\boldsymbol{u}$, then the mapping $E \mapsto D$ is an LTI system with frequency response $\epsilon(\boldsymbol{x}, -\cdot)$. Because the electric displacement in a medium depends only on the electric field in the past, the system is causal, i.e., $\operatorname{supp} \widehat{\epsilon(\boldsymbol{x}, \cdot)} \subset [0, \infty)$.

Example 2.16 (Reflection coefficient). *We consider a situation as described in Chapter 1, where an electromagnetic field is incident perpendicularly on a layered medium. We assume that the layered medium is surrounded by air with refractive index $n_0 = 1$. The incident field is a plane wave coming from the left, i.e., $\boldsymbol{\mathcal{E}}(\boldsymbol{x}, t) = E_{\text{in}}(x)\mathrm{e}^{-i\omega t}\boldsymbol{e}_y$ with $E_{\text{in}}(x) = \mathrm{e}^{ikx}$. The layered medium gives rise to a reflected field $\boldsymbol{\mathcal{E}}(\boldsymbol{x}, t) = E_{\text{ref}}(x)\mathrm{e}^{-i\omega t}\boldsymbol{e}_y$ with $E_{\text{ref}}(x) = R(k)\mathrm{e}^{-ikx}$. The number $R(k)$ is called* reflection coefficient *(from the left). The reflected field is also a plane wave, but it travels in the opposite direction.*

Then the mapping $L : E_{\text{in}} \mapsto \overline{E_{\text{ref}}}$ is an LTI system which maps $\mathrm{e}^{ik\cdot}$ to $\overline{R(k)}\mathrm{e}^{ik\cdot}$. Because the system should map a real incident field to a real reflected field, its impulse response should only take values in $\mathbb{R}$. Therefore, $\overline{R(k)} = R(-k)$, i.e., the frequency response of L is R°. Because the layered structure cannot reflect a field before an incident field has arrived, the system is causal, i.e., $\operatorname{supp} \widehat{R} \subset [0, \infty)$.

We make all these statements rigorous in Chapter 3.

2.3 The Paley-Wiener Theorem

In this section we prove a generalized version of the Paley-Wiener Theorem.

Theorem 2.17 (Paley-Wiener). *Let $1 \leq p \leq \infty$. We have the characterization*

$$H^p(\mathbb{C}^+) = \left\{ f \in L^p(\mathbb{R}) : \operatorname{supp} \widehat{f} \subset [0, \infty) \right\}.$$

The theorem follows directly from Propositions 2.19 and 2.20 below. We begin with a lemma.

Lemma 2.18. *$H^\infty(\mathbb{C}^+)$ is (sequentially) weak*-closed in $L^\infty(\mathbb{R})$.*

Proof. Let $(f_n) \subset H^\infty(\mathbb{C}^+)$ with $f_n \stackrel{*}{\rightharpoonup} f$ for some $f \in L^\infty(\mathbb{R})$, i.e., $\int_{\mathbb{R}} (f_n - f)\varphi \to 0$ as $n \to \infty$ for all $\varphi \in L^1(\mathbb{R})$. We need to show that $f \in H^\infty(\mathbb{C}^+)$. To this end we define a function F on the upper half-plane by

$$F(x+iy) = \frac{1}{\pi} \int_{\mathbb{R}} f(t) \frac{y}{(x-t)^2 + y^2} \, \mathrm{d}t$$

for real x and $y > 0$. By Theorem 2.10, F is bounded on $\mathbb{C}^+$ and has boundary values f on the real line. In the following, we identify F with f, that is, we denote by f the original function in $L^p(\mathbb{R})$ as well as the function on the upper half-plane defined by the above integral formula. It remains to show that f is analytic on $\mathbb{C}^+$.

We are going to show that f_n converges to f uniformly on compact subsets of $\mathbb{C}^+$. So let $K \subset \mathbb{C}^+$ be compact and fix an arbitrary $\epsilon > 0$. We use the abbreviation $P_{x,y}(t) = \frac{y}{\pi((x-t)^2+y^2)}$. Choose finitely many points $z_j = x_j + iy_j \in K$ such that for every $z = x + iy \in K$ there is a j with $\|P_{x,y} - P_{x_j,y_j}\|_{L^1(\mathbb{R})} < \frac{\epsilon}{2\sup_{n\in\mathbb{N}} \|f_n - f\|_{L^\infty(\mathbb{R})}}$. Notice that the supremum on the right hand side is finite since weakly*-convergent series are bounded. Finally, choose $M > 0$ so large that $\left|\int_{\mathbb{R}} (f_n - f) P_{x_j,y_j}\right| < \frac{\epsilon}{2}$ for all j and for all $n > M$. This is possible since $f_n \stackrel{*}{\rightharpoonup} f$.

Now let $z = x + iy \in K$ be arbitrary. Pick some j such that $\|P_{x,y} - P_{x_j,y_j}\|_{L^1(\mathbb{R})} < \frac{\epsilon}{2\sup_{n\in\mathbb{N}} \|f_n - f\|_{L^\infty(\mathbb{R})}}$. We have the representation

$$f_n(x+iy) = \frac{1}{\pi} \int_{\mathbb{R}} f_n(t) \frac{y^2}{(x-t)^2 + y^2} \, \mathrm{d}t$$

by Theorem 2.11. It follows that for $n > M$

$$\begin{aligned}
|f(x+iy) - f_n(x+iy)| &= \left| \int_{\mathbb{R}} (f(t) - f_n(t))\, P_{x,y}(t)\, \mathrm{d}t \right| \\
&\leq \left| \int_{\mathbb{R}} (f(t) - f_n(t)) \left(P_{x,y}(t) - P_{x_j,y_j}(t) \right) \mathrm{d}t \right| \\
&\qquad + \left| \int_{\mathbb{R}} (f(t) - f_n(t))\, P_{x_j,y_j}(t)\, \mathrm{d}t \right| \\
&\leq \|f_n - f\|_{L^\infty(\mathbb{R})} \|P_{x,y} - P_{x_j,y_j}\|_{L^1(\mathbb{R})} + \frac{\epsilon}{2} \\
&\leq \frac{\epsilon}{2} + \frac{\epsilon}{2} = \epsilon.
\end{aligned}$$

Thus, f_n converges uniformly to f on compact subsets of $\mathbb{C}^+$. Since the f_n are analytic, f is also analytic on $\mathbb{C}^+$.

To summarize, we have shown that f is bounded and analytic on the upper half-plane. This means that $f \in H^\infty(\mathbb{C}^+)$. □

We can now prove one direction of a generalized Paley-Wiener theorem.

Proposition 2.19. *Let* $1 \leq p \leq \infty$ *and* $f \in L^p(\mathbb{R})$ *with* $\operatorname{supp} \widehat{f} \subset [0, \infty)$. *Then* $f \in H^p(\mathbb{C}^+)$.

Proof. We begin with the case $p = 1$. So let $f \in L^1(\mathbb{R})$ with $\operatorname{supp} \widehat{f} \subset [0, \infty)$. Define a function F on $\mathbb{C}^+$ by

$$F(z) = \frac{1}{2\pi} \int_{\mathbb{R}} \widehat{f}(\xi) \mathrm{e}^{iz\xi}\, \mathrm{d}\xi.$$

We are going to show that $F \in H^1(\mathbb{C}^+)$ and that F has boundary values f on the real line. Notice two things: First of all, the integral converges since $\operatorname{supp} \widehat{f} \subset [0, \infty)$ and $\mathrm{e}^{iz\xi}$ decays exponentially as $\xi \to \infty$ for $z \in \mathbb{C}^+$. Especially, we can differentiate under the integral sign, whence F is analytic on $\mathbb{C}^+$. Second, on the real line the above formula is formally the well-known Fourier inversion formula. Therefore, it seems reasonable that F has boundary values f on the real line. Let us make this rigorous.

For $y > 0$ write $F_y(x) = F(x + iy)$. Then

$$F_y(x) = \frac{1}{2\pi} \int_{\mathbb{R}} \widehat{f}(\xi) H(\xi) \mathrm{e}^{-y\xi} \mathrm{e}^{ix\xi}\, \mathrm{d}\xi,$$

where H is the Heaviside function. Denoting by $\widetilde{g}$ the inverse Fourier transform of a function g,

$$F_y(x) = \left(\xi \mapsto \widehat{f}(\xi)H(\xi)\mathrm{e}^{-y\xi}\right)^{\sim}(x) = \left(f * \left(\xi \mapsto H(\xi)\mathrm{e}^{-y\xi}\right)^{\sim}\right)(x)$$
$$= \frac{1}{2\pi}\left(f * \left(t \mapsto \frac{1}{y - it}\right)\right)(x) = \frac{1}{2\pi}\int_{\mathbb{R}} f(s)\frac{1}{y - i(x-s)}\,\mathrm{d}s.$$

On the other hand, with the abbreviation $Q_y(t) = \frac{1}{y+it}$,

$$\int_{\mathbb{R}} f(s)\frac{1}{y + i(x-s)}\,\mathrm{d}s = (f * Q_y)(x) = \left((\widehat{f})^{\sim} * (\widehat{Q_y})^{\sim}\right)(x)$$
$$= \left(\widehat{f}\widehat{Q_y}\right)^{\sim}(x) = 0,$$

since $\operatorname{supp}\widehat{f} \subset [0,\infty)$, while $\widehat{Q_y}(\tau) = \mathrm{e}^{y\tau}H(-\tau)$ is supported in $(-\infty, 0]$. It follows that

$$F_y(x) = \frac{1}{2\pi}\int_{\mathbb{R}} f(s)\left(\frac{1}{y - i(x-s)} + \frac{1}{y + i(x-s)}\right)\mathrm{d}s$$
$$= \frac{1}{\pi}\int_{\mathbb{R}} f(s)\frac{y}{(x-s)^2 + y^2}\,\mathrm{d}s.$$

This is a good thing. By Theorem 2.10 this implies that $\sup_{y>0}\|F_y\|_{L^1(\mathbb{R})} < \infty$ and that F has boundary values f on the real line. Since F is also analytic on $\mathbb{C}^+$, $F \in H^1(\mathbb{C}^+)$. Of course, we identify f with F, i.e., $f \in H^1(\mathbb{C}^+)$.

Now consider the case $1 < p \leq \infty$. Let $f \in L^p(\mathbb{R})$ with $\operatorname{supp}\widehat{f} \subset [0,\infty)$. We use approximation and reduce to the cases $p = 1$ or $p = 2$. Define

$$f_n(z) = \gamma_n(z)f(z),$$

where

$$\gamma_n(z) = 1 - \left(\frac{-iz}{-iz+1}\right)^n.$$

Since $\left|1 - \left(\frac{-iz}{-iz+1}\right)^n\right| = \left|\frac{(-iz+1)^n - (-iz)^n}{(-iz+1)^n}\right| = O\left(\frac{1}{|z|}\right)$ as $|z| \to \infty$, it follows that γ_n is in $H^p(\mathbb{C}^+)$ for every $1 < p \leq \infty$. For $1 < p < \infty$, Hölder's inequality shows that $f_n \in L^1(\mathbb{R})$:

$$\|f_n\|_{L^1(\mathbb{R})} \leq \|\gamma_n\|_{L^q(\mathbb{R})}\|f\|_{L^p(\mathbb{R})} < \infty,$$

where $\frac{1}{p} + \frac{1}{q} = 1$. For $p = \infty$ a similar estimate yields $f_n \in L^2(\mathbb{R})$. A direct calculation moreover shows that $\operatorname{supp}\widehat{\gamma_n} \subset [0,\infty)$. This implies that

$\widehat{f_n} = \frac{1}{2\pi}\widehat{\gamma_n} * \widehat{f}$ is also supported in $[0,\infty)$. Hence, for $p=\infty$, $f_n \in H^2(\mathbb{C}^+)$ by the classical Paley-Wiener theorem, and for $1 < p < \infty$, $f_n \in H^1(\mathbb{C}^+)$ by what we have already shown. By Theorem 2.11, this implies especially that in $\mathbb{C}^+$ we have the representation

$$f_n(x+iy) = \frac{1}{\pi}\int_{\mathbb{R}} f_n(t) P_y(x-t)\,\mathrm{d}t,$$

where $P_y(x) = \frac{y}{x^2+y^2}$. By Theorem 2.10, the L^p-norms of f_n along lines parallel to $\mathbb{R}$ in the upper half-plane $\mathbb{C}^+$ are bounded, whence $f_n \in H^p(\mathbb{C}^+)$.

It is now easy to check that $\left|\frac{-iz}{-iz+1}\right| < 1$ for $z \in \overline{\mathbb{C}^+}$, so that $\left|\frac{-iz}{-iz+1}\right|^n \to 0$ pointwise for every fixed $z \in \overline{\mathbb{C}^+}$ as $n \to \infty$. Thus, for $1 < p < \infty$, $f_n \to f$ in $L^p(\mathbb{R})$ by dominated convergence. Since $H^p(\mathbb{C}^+)$ is a closed subspace of $L^p(\mathbb{R})$, $f \in H^p(\mathbb{C}^+)$. For $p = \infty$, we have for any $\varphi \in L^1(\mathbb{R})$

$$\left|\int_{\mathbb{R}} (f - f_n)(z)\varphi(z)\,\mathrm{d}z\right| = \left|\int_{\mathbb{R}} \left(\frac{-iz}{-iz+1}\right)^n f(z)\varphi(z)\,\mathrm{d}z\right| \to 0 \qquad (2.3)$$

as $n \to \infty$ by dominated convergence. So $f_n \xrightarrow{*} f$ in $L^\infty(\mathbb{R})$. Since $H^\infty(\mathbb{C}^+)$ is sequentially weak*-closed in $L^\infty(\mathbb{R})$ by Lemma 2.18, $f \in H^\infty(\mathbb{C}^+)$. □

We now prove the converse of Proposition 2.19.

Proposition 2.20. *Let* $1 \le p \le \infty$ *and* $f \in H^p(\mathbb{C}^+)$. *Then* $\operatorname{supp}\widehat{f} \subset [0,\infty)$.

Proof. For $1 \le p \le 2$, the theorem is well-known. A proof for $f \in H^1(\mathbb{C}^+)$ and $f \in H^2(\mathbb{C}^+)$ can for example be found in [32, Chapter 8]. For the general case, define as in the proof of Proposition 2.19

$$f_n(z) = \gamma_n(z) f(z),$$

where

$$\gamma_n(z) = 1 - \left(\frac{-iz}{-iz+1}\right)^n.$$

As before, Hölder's inequality yields that $f_n \in H^1(\mathbb{C}^+)$ for $p < \infty$ and $f_n \in H^2(\mathbb{C}^+)$ for $p = \infty$. As already mentioned, it is well-known that this implies $\operatorname{supp}\widehat{f_n} \subset [0,\infty)$.

Now fix $\varphi \in \mathcal{S}(\mathbb{R})$, the Schwartz class. Like in (2.3) we get that $\int_{\mathbb{R}} f_n\varphi \to \int_{\mathbb{R}} f\varphi$ as $n \to \infty$. This means that $f_n \to f$ in $\mathcal{S}'(\mathbb{R})$. Since the Fourier transform is continuous as a map from $\mathcal{S}'(\mathbb{R})$ to $\mathcal{S}'(\mathbb{R})$ and since $\operatorname{supp}\widehat{f_n} \subset [0,\infty)$, it follows that also $\operatorname{supp}\widehat{f} \subset [0,\infty)$. □

Chapter 3

Scattering Theory for the 1D Helmholtz Equation

A typical example of an inverse problem is the following. Given the following answer: The Answer to the Great Question of Life, the Universe and Everything is forty-two, find the question.

from http://www-sop.inria.fr/apics/research.html

In this chapter we study the *reflection coefficient* and the *transmission coefficient* for the one-dimensional Helmholtz equation

$$u''(x) + k^2 n^2(x) u(x) = 0, \tag{3.1}$$

which we derived in Chapter 1 as a model for the propagation of electromagnetic waves in layered media. We assume throughout this chapter that the layered medium itself is *non-dispersive*, i.e., that n does not depend on k, and *non-absorbing*, i.e., that n is real-valued. This approximation is sufficient for our theoretical investigations, because the refractive index of materials that are typically used for optical interference coatings varies only slightly for optical frequencies. Even in the actual design process the frequency-dependency of the refractive index of the coating materials is often neglected until the final optimization step [40].

Throughout most of this chapter we model the situation of a layered structure that is surrounded by air or vacuum, i.e., $n(x) = 1$ for $x \notin [0, d]$ with some $d > 0$. Further, only a certain range of refractive indices can be physically achieved. Therefore, we assume that $n|_{[0,d]} \in L^\infty_{a,b}(0, d)$ for some $0 < a < b$, where

$$L^\infty_{a,b}(0, d) = \{f \in L^\infty(0, d) : a \le f(x) \le b \text{ for almost all } x \in [0, d]\}.$$

We also write $n \in L^\infty_{a,b}(0,d)$ when we mean $n|_{[0,d]} \in L^\infty_{a,b}(0,d)$ and $n|_{\mathbb{R}\setminus[0,d]} = 1$.

We are only concerned with the *direct scattering problem*, i.e., for a given refractive profile n, we study the properties of the corresponding reflection coefficients from the right and from the left, R_1 and R_2, and the transmission coefficient T. In Section 3.1 we define reflection and transmission coefficient for the Helmholtz equation and prove some of their basic properties. In Section 3.2 we show that reflection and transmission coefficient satisfy a continuity property with respect to the refractive profile. A continuity property is also used in Section 3.3 to prove Hardy space properties of reflection and transmission coefficient. As another application of the results from Section 3.2 we show in Section 3.4 that a certain optimization problem for the reflection coefficient has a solution. We finish with some remarks on what is known if the refractive profile n is smooth in Section 3.5.

3.1 The direct scattering problem

The material in this section is rather technical, but our approach in Sections 3.1.1–3.1.3 is quite standard. Analogous considerations for the Schrödinger equation can for example be found in [18]. The main result of this section is Theorem 3.5. Parts (a) and (b) of Theorem 3.5 are already in [28]. For the sake of completeness and because the notation in [28] differs very much from our notation, we prove those parts anyway. Concerning the rest of the theorems and formulas in this section, we are not aware of any reference where they are explicitly stated.

3.1.1 Jost solutions and an integral formulation

In order to define reflection and transmission coefficient for the Helmholtz equation, we need to consider solutions of (3.1) that represent incident plane waves. Let $u_1(x,k)$ be the solution of (3.1) with initial conditions

$$\begin{cases} u_1(d,k) = e^{ikd}, \\ u_1'(d,k) = ike^{ikd}, \end{cases} \tag{3.2}$$

and let $u_2(x,k)$ be the solution of (3.1) with initial conditions

$$\begin{cases} u_2(0,k) = 1, \\ u_2'(0,k) = -ik. \end{cases} \tag{3.3}$$

u_1 and u_2 are called the *Jost solutions* of (3.1). Obviously, we have $u_1(x,k) = \mathrm{e}^{ikx}$ for $x \geq d$ and $u_2(x,k) = \mathrm{e}^{-ikx}$ for $x \leq 0$. As usual, it is easier to obtain properties of u_1 and u_2 by considering integral equations that are equivalent to the above initial value problems. To this end, we define the *Faddeev functions*

$$\begin{aligned} m_1(x,k) &= \mathrm{e}^{-ikx} u_1(x,k), \\ m_2(x,k) &= \mathrm{e}^{ikx} u_2(x,k). \end{aligned}$$

Obviously, we have $m_1(x,k) = 1$, $x \geq d$, and $m_2(x,k) = 1$, $x \leq 0$. One easily checks that m_1 solves

$$\begin{cases} m_1''(x,k) + 2ikm_1'(x,k) + k^2(n^2(x) - 1)m_1(x,k) = 0, \\ m_1(d,k) = 1, \\ m_1'(d,k) = 0, \end{cases}$$

while m_2 solves

$$\begin{cases} m_2''(x,k) - 2ikm_2'(x,k) + k^2(n^2(x) - 1)m_2(x,k) = 0, \\ m_2(0,k) = 1, \\ m_2'(0,k) = 0. \end{cases}$$

The derivatives are of course taken with respect to x. Variation of constants then leads to the equivalent integral equations

$$m_1(x,k) = 1 + \int_x^d m_1(t,k)\,\big((1 - n^2(t))D_k(t - x)\big)\,\mathrm{d}t \tag{3.4}$$

and

$$m_2(x,k) = 1 + \int_0^x m_2(t,k)\,\big((1 - n^2(t))D_k(x - t)\big)\,\mathrm{d}t, \tag{3.5}$$

where

$$D_k(y) = k\frac{\mathrm{e}^{2iky} - 1}{2i}.$$

3.1.2 Estimates for Jost solutions

The following lemma gives a bound on the solution of certain integral equations of Volterra type.

Lemma 3.1. *Consider the Volterra integral equation*

$$u(x) = g(x) + \int_0^x K(x,t)u(t)\,\mathrm{d}t, \tag{3.6}$$

where $g \in L^\infty_{\mathrm{loc}}(\mathbb{R})$ and there is $h \in L^1_{\mathrm{loc}}(\mathbb{R})$ such that $|K(x,\cdot)| \le h$ for almost all $x \in \mathbb{R}$. This integral equation has a unique solution u, and for $x \ge 0$ we have the estimate

$$|u(x)| \le \left(\sup_{t\in[0,x]} |g(t)|\right) \exp\left(\int_0^x h(t)\,\mathrm{d}t\right).$$

Proof. Let $Au(x) = \int_0^x K(x,t)u(t)\,\mathrm{d}t$, and let $x_0 > 0$. Then A is a bounded linear operator on $L^\infty([0,x_0])$. It is not hard to show that for $j \in \mathbb{N}$

$$A^j u(x) = \int_{0\le x_1\le\ldots\le x_j\le x} K(x_2,x_1)\ldots K(x_j,x_{j-1})K(x,x_j)u(x_1)\,\mathrm{d}x_1\ldots\,\mathrm{d}x_j,$$

and then

$$\begin{aligned}|A^j u(x)| &\le \left(\sup_{t\in[0,x]} |u(t)|\right) \int_{0\le x_1\le\ldots\le x_j\le x} h(x_1)\ldots h(x_j)\,\mathrm{d}x_1\ldots\,\mathrm{d}x_j \\ &= \left(\sup_{t\in[0,x]} |u(t)|\right) \frac{\left(\int_0^x h(t)\,\mathrm{d}t\right)^j}{j!}.\end{aligned}$$

So there is $j \in \mathbb{N}$ with $\|A^j\|_{\mathcal{L}(L^\infty([0,x_0]))} < 1$. It follows that $I - A$ is invertible and $(I-A)^{-1} = \sum_{j=0}^\infty A^j$ with convergence in $\mathcal{L}(L^\infty([0,x_0]))$. In fact, this is still true if we consider A as an operator on $L^\infty([-x_0,x_0])$. Applying this to the integral equation (3.6) yields the unique solution

$$u(x) = \sum_{j=0}^\infty (A^j g)(x), \tag{3.7}$$

where convergence is locally uniform. Moreover,

$$|u(x)| \le \left(\sup_{t\in[0,x]} |g(t)|\right) \sum_{j=0}^\infty \frac{\left(\int_0^x h(t)\,\mathrm{d}t\right)^j}{j!} = \left(\sup_{t\in[0,x]} |g(t)|\right) \exp\left(\int_0^x h(t)\,\mathrm{d}t\right).$$

□

We can use Lemma 3.1 to obtain estimates for the solutions of (3.4) and (3.5). By $\dot{m}_j$, $j = 1, 2$, we denote the derivative of m_j with respect to k.

Theorem 3.2. *Equations* (3.4) *and* (3.5) *for* m_1 *and* m_2 *are uniquely solvable for each* $k \in \mathbb{C}$. *The solutions can be bounded as follows:*

(a) *For* $\operatorname{Im} k \geq 0$ *and* $x \leq d$

$$|m_1(x,k)| \leq \exp\left(|k| \int_x^d |1 - n^2(t)| \,\mathrm{d}t\right).$$

(b) *For* $\operatorname{Im} k \geq 0$ *and* $x \leq d$

$$|\dot{m}_1(x,k)| \leq \left((d-x) + \frac{|k|}{2}(d-x)^2\right)\left(\sup_{t\in[x,d]} |1 - n^2(t)|\right) \cdot \exp\left(2|k| \int_x^d |1 - n^2(t)| \,\mathrm{d}t\right).$$

(c) *For* $\operatorname{Im} k \geq 0$ *and* $x \geq 0$

$$|m_2(x,k)| \leq \exp\left(|k| \int_0^x |1 - n^2(t)| \,\mathrm{d}t\right).$$

(d) *For* $\operatorname{Im} k \geq 0$ *and* $x \geq 0$

$$|\dot{m}_2(x,k)| \leq \left(x + \frac{|k|}{2}x^2\right)\left(\sup_{t\in[0,x]} |1 - n^2(t)|\right) \exp\left(2|k| \int_0^x |1 - n^2(t)| \,\mathrm{d}t\right).$$

Proof. We only prove the estimates for m_2. Applying Lemma 3.1 to (3.5) and using

$$|D_k(y)| \leq |k| \tag{3.8}$$

for $\operatorname{Im} k \geq 0$ and $y \in \mathbb{R}$ already yields (c).

To obtain the estimate for $\dot{m}_2$ we are going to apply Lemma 3.1 to

$$\begin{aligned} \dot{m}_2(x,k) = &\int_0^x \dot{m}_2(t,k)(1 - n^2(t)) D_k(x-t) \,\mathrm{d}t \\ &+ \int_0^x m_2(t,k)(1 - n^2(t)) \dot{D}_k(x-t) \,\mathrm{d}t. \end{aligned} \tag{3.9}$$

The above equation follows from (3.5). First note that

$$|\dot{D}_k(y)| = \left| \frac{e^{2iky} - 1}{2i} + kye^{2iky} \right| \leq 1 + |k|y$$

for $\operatorname{Im} k \geq 0$ and $y \geq 0$. Together with (c) we get for $x \geq 0$

$$\begin{aligned} &\left| \int_0^x m_2(t,k)(1-n^2(t))\dot{D}_k(x-t)\,\mathrm{d}t \right| \\ &\quad \leq \int_0^x \exp\left(|k| \int_0^t |1-n^2(s)|\,\mathrm{d}s \right) |1-n^2(t)|\,(1+|k|(x-t))\,\mathrm{d}t \\ &\quad \leq \left(x + \frac{|k|}{2}x^2 \right) \left(\sup_{t\in[0,x]} |1-n^2(t)| \right) \exp\left(|k| \int_0^x |1-n^2(t)|\,\mathrm{d}t \right). \end{aligned} \tag{3.10}$$

Now (d) follows by applying Lemma 3.1 to (3.9) and using (3.8) and (3.10).
The estimates for m_1 can be obtained in the same way. □

Remark 3.3. *When one inspects the series (3.7) for m_1 and m_2, one sees easily that it converges uniformly in k on compact subsets of $\mathbb{C}$. Since the partial sums themselves are analytic in $\mathbb{C}$, $m_1(x,\cdot)$ and $m_2(x,\cdot)$ are analytic in $\mathbb{C}$.*

Remark 3.4. *One easily sees that since n is real-valued, the symmetry relations (or* reality conditions*) $m_j(x,-\overline{k}) = \overline{m_j(x,k)}$ and $u_j(x,-\overline{k}) = \overline{u_j(x,k)}$, $j = 1,2$, hold.*

3.1.3 Reflection and transmission coefficient R and T

For $k \in \mathbb{C} \setminus \{0\}$ the Jost solutions $u_2(\cdot,k)$ and $u_2(\cdot,-k)$ are linearly independent. This can be seen as follows: The Wronskian $W[u_2(x,k),u_2(x,-k)]$ does not depend on x since there appears no first derivative in (3.1) (compare [65, §15.III and §19]). Since

$$\begin{aligned} W[u_2(x,k),u_2(x,-k)] &= u_2(x,k)u_2'(x,-k) - u_2'(x,k)u_2(x,-k) \\ &= u_2(0,k)u_2'(0,-k) - u_2'(0,k)u_2(0,-k) \\ &= 2ik \neq 0, \end{aligned}$$

$u_2(\cdot,k)$ and $u_2(\cdot,-k)$ are linearly independent. Similarly,

$$W[u_1(x,k),u_1(x,-k)] = -2ik \neq 0, \tag{3.11}$$

whence $u_1(\cdot, k)$ and $u_1(\cdot, -k)$ are linearly independent. It follows that there are functions $\alpha(k)$, $\beta(k)$, $\gamma(k)$ and $\delta(k)$ such that

$$u_2(x,k) = \alpha(k)u_1(x,k) + \beta(k)u_1(x,-k), \tag{3.12}$$
$$u_1(x,k) = \gamma(k)u_2(x,k) + \delta(k)u_2(x,-k) \tag{3.13}$$

for $k \neq 0$. We then define

$$T_1(k) = \frac{1}{\beta(k)}, \qquad R_1(k) = \frac{\alpha(k)}{\beta(k)},$$
$$T_2(k) = \frac{1}{\delta(k)}, \qquad R_2(k) = \frac{\gamma(k)}{\delta(k)}.$$

The functions T_1 and R_1 are called *transmission coefficient from the right* and *reflection coefficient from the right*, respectively, and T_2 and R_2 are called *transmission coefficient from the left* and *reflection coefficient from the left*, respectively. The reason for these definitions is as follows. After multiplying (3.12) and (3.13) by $T_1(k)$ and $T_2(k)$, respectively, we obtain

$$T_1(k)u_2(x,k) = R_1(k)u_1(x,k) + u_1(x,-k), \tag{3.14}$$
$$T_2(k)u_1(x,k) = R_2(k)u_2(x,k) + u_2(x,-k). \tag{3.15}$$

Let us consider the second equation. Fix some $k > 0$. Then

$$u(x) = T_2(k)u_1(x,k) = R_2(k)u_2(x,k) + u_2(x,-k)$$

solves the Helmholtz equation (3.1). For $x \leq 0$ we have

$$u(x) = R_2(k)u_2(x,k) + u_2(x,-k) = R_2(k)\mathrm{e}^{-ikx} + \mathrm{e}^{ikx},$$

so to the left of the layered medium u is a plane wave travelling to the right plus a plane wave with complex amplitude $R_2(k)$ travelling to the left. For $x \geq d$ we have

$$u(x) = T_2(k)u_1(x,k) = T_2(k)\mathrm{e}^{ikx},$$

so to the right of the layered medium u is a plane wave with complex amplitude $T_2(k)$ travelling to the right, see Figure 3.1. To sum up, an incoming plane wave from the left, $u_{\text{in}}(x) = \mathrm{e}^{ikx}$, gives rise to a transmitted wave $u_{\text{trans}}(x) = T_2(x)\mathrm{e}^{ikx}$ and a reflected wave travelling in the opposite direction, $u_{\text{ref}}(x) = R_2(k)\mathrm{e}^{-ikx}$. Similar considerations for the second equation explain the definition of T_1 and R_1.

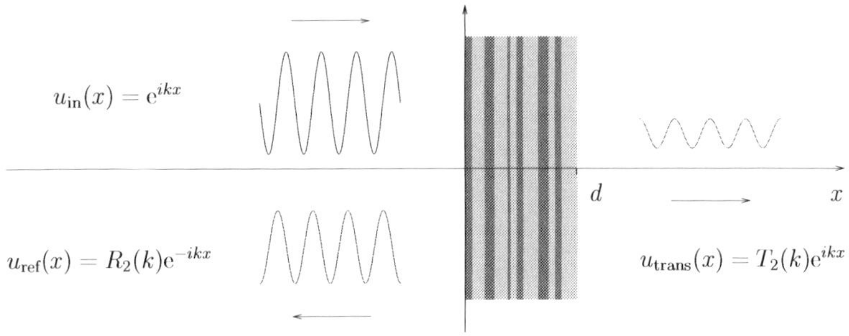

Figure 3.1: When a plane wave $u_{\text{in}}(x) = e^{ikx}$ is incident on a layered structure, part of it is reflected ($u_{\text{ref}}(x) = R_2(k)e^{-ikx}$) and part of it is transmitted ($u_{\text{trans}}(x) = T_2(k)e^{ikx}$).

In the following it will also turn out useful to rewrite (3.12) and (3.13) in terms of m_1 and m_2,

$$T_1(k)m_2(x,k) = R_1(k)e^{2ikx}m_1(x,k) + m_1(x,-k), \tag{3.16}$$
$$T_2(k)m_1(x,k) = R_2(k)e^{-2ikx}m_2(x,k) + m_2(x,-k). \tag{3.17}$$

Of course, it is not quite clear yet that the reflection and transmission coefficients are defined everywhere, since β or δ might have zeros. Also, it is not clear yet what happens at $k = 0$. The following theorem sheds some light on this.

Theorem 3.5. (a) *α, β, γ and δ can be extended to functions analytic on $\mathbb{C}$. Especially, the transmission coefficients $T_1 = \frac{1}{\beta}$, $T_2 = \frac{1}{\delta}$ have no zeros in $\mathbb{C}$. R_1, R_2, T_1 and T_2 are functions meromorphic on $\mathbb{C}$ and holomorphic on the closed upper half-plane $\overline{\mathbb{C}^+} = \{\operatorname{Im} z \geq 0\}$. Especially, they are C^∞-functions on the real line.*

(b) *We have the following relations:*

(i) *For $k \in \mathbb{C}$, $T(k) := T_1(k) = T_2(k)$.*

(ii) *For $k \in \mathbb{C}$, the* reality conditions *$\overline{T(k)} = T(-\overline{k})$, $\overline{R_1(k)} = R_1(-\overline{k})$, $\overline{R_2(k)} = R_2(-\overline{k})$ hold.*

(iii) *Conservation of energy: For $k \in \mathbb{R}$, $|T(k)|^2 + |R_1(k)|^2 = |T(k)|^2 + |R_2(k)|^2 = 1$.*

(c) *The following integral representations hold:*

$$\begin{aligned}
\frac{R_1(k)}{T(k)} &= \frac{1}{2i}k\int_0^d \mathrm{e}^{-2ikt}(1-n^2(t))m_2(t,k)\,\mathrm{d}t,\\
\frac{R_2(k)}{T(k)} &= \frac{1}{2i}k\int_0^d \mathrm{e}^{2ikt}(1-n^2(t))m_1(t,k)\,\mathrm{d}t,\\
\frac{1}{T(k)} &= 1-\frac{1}{2i}k\int_0^d (1-n^2(t))m_1(t,k)\,\mathrm{d}t\\
&= 1-\frac{1}{2i}k\int_0^d (1-n^2(t))m_2(t,k)\,\mathrm{d}t.
\end{aligned}$$

Analyticity on the complex upper half-plane is also called the *causality condition.* The careful reader of Chapter 2 will immediately understand this notion.

Proof. Since the quantities that we are interested in are the reflection and transmission coefficients, we write $\frac{R_1(k)}{T_1(k)}$, $\frac{1}{T_1(k)}$ and so on instead of $\alpha(k)$, $\beta(k)$ and so on. We begin with the proof of (b). Observe that

$$\begin{aligned}
W[u_1(x,k),u_2(x,k)] &\overset{(3.12)}{=} W[u_1(x,k),\frac{R_1(k)}{T_1(k)}u_1(x,k)+\frac{1}{T_1(k)}u_1(x,-k)]\\
&= \frac{1}{T_1(k)}W[u_1(x,k),u_1(x,-k)]\\
&= -2ik\frac{1}{T_1(k)}.
\end{aligned} \tag{3.18}$$

On the other hand, using (3.13) instead, we obtain

$$W[u_1(x,k),u_2(x,k)] = -2ik\frac{1}{T_2(k)}. \tag{3.19}$$

This implies, at least for $k \in \mathbb{C}\setminus\{0\}$, $T_1(k) = T_2(k)$, i.e., (i). In the following, we simply write $T(k)$.

Similarly, one can compute

$$2ik\frac{R_1(k)}{T(k)} = W[u_1(x,-k),u_2(x,k)], \tag{3.20}$$

$$2ik\frac{R_2(k)}{T(k)} = W[u_1(x,k),u_2(x,-k)]. \tag{3.21}$$

Furthermore, using (3.18) and Remark 3.4,

$$\begin{aligned}\frac{1}{\overline{T(k)}} &= \frac{1}{-2i(-\bar{k})} W[\overline{u_1(x,k)}, \overline{u_2(x,k)}] \\ &= \frac{1}{-2i(-\bar{k})} W[u_1(x,-\bar{k}), u_2(x,-\bar{k})] = \frac{1}{T(-\bar{k})}.\end{aligned}$$

Using (3.20) or (3.21) instead, we get $\overline{R_1(k)} = R_1(-\bar{k})$ and $\overline{R_2(k)} = R_2(-\bar{k})$, respectively. This is (ii).

By (3.20), (3.21) and Remark 3.4,

$$\frac{R_1(k)}{T(k)} = -\frac{R_2(-k)}{T(-k)}. \tag{3.22}$$

Plugging (3.12) into (3.13),

$$\begin{aligned} u_1(x,k) &= \frac{R_1(k)R_2(k)}{T(k)^2} u_1(x,k) + \frac{R_2(k)}{T(k)^2} u_1(x,-k) \\ &\quad + \frac{R_1(-k)}{T(k)T(-k)} u_1(x,-k) + \frac{1}{T(k)T(-k)} u_1(x,k) \\ &\overset{(3.22)}{=} \frac{-R_1(k)R_1(-k)}{T(k)T(-k)} u_1(x,k) + \frac{-R_1(-k)}{T(k)T(-k)} u_1(x,-k) \\ &\quad + \frac{R_1(-k)}{T(k)T(-k)} u_1(x,-k) + \frac{1}{T(k)T(-k)} u_1(x,k) \\ &= \frac{1 - R_1(k)R_1(-k)}{T(k)T(-k)} u_1(x,k). \end{aligned}$$

Because of (ii) this implies $|R_1(k)|^2 + |T(k)|^2 = 1$ for $k \in \mathbb{R} \setminus \{0\}$. The relation $|R_2(k)|^2 + |T(k)|^2 = 1$ for $k \in \mathbb{R} \setminus \{0\}$ is derived in the same way. This finishes the proof of (b).

Let us now prove (c). For $x \leq 0$ we have by (3.4) (and since $n(x) = 1$ for $x \leq 0$)

$$\begin{aligned} m_1(x,k) &= 1 + \int_0^d m_1(t,k)\left(1 - n^2(t)\right) k \frac{\mathrm{e}^{2ik(t-x)} - 1}{2i}\,\mathrm{d}t \\ &= 1 - \frac{k}{2i}\int_0^d m_1(t,k)\left(1 - n^2(t)\right)\mathrm{d}t \\ &\quad + \left(\frac{k}{2i}\int_0^d m_1(t,k)\left(1 - n^2(t)\right)\mathrm{e}^{2ikt}\,\mathrm{d}t\right)\mathrm{e}^{-2ikx}. \end{aligned}$$

On the other hand, by (3.17) we have for $x \leq 0$

$$m_1(x,k) = \frac{R_2(k)}{T(k)} \mathrm{e}^{-2ikx} m_2(x,k) + \frac{1}{T(k)} m_2(x,-k) = \frac{R_2(k)}{T(k)} \mathrm{e}^{-2ikx} + \frac{1}{T(k)}.$$

Together, it follows that

$$\frac{R_2(k)}{T(k)} = \frac{k}{2i} \int_0^d m_1(t,k) \left(1 - n^2(t)\right) \mathrm{e}^{2ikt} \, \mathrm{d}t, \tag{3.23}$$

$$\frac{1}{T(k)} = 1 - \frac{k}{2i} \int_0^d m_1(t,k) \left(1 - n^2(t)\right) \mathrm{d}t, \tag{3.24}$$

which are two of the formulas of (c). The other two formulas of (c) are obtained in the same way by considering the integral equation for m_2 instead.

We remark that from the integral representations we especially get $T(0) = 1$ and $R_1(0) = R_2(0) = 0$, so the formulas in (b) also hold for $k = 0$ (which we had not treated in this proof so far).

It remains to show (a). For example, from the integral representation (3.24) we see that $\beta(k) = \frac{1}{T(k)}$ is analytic in $\mathbb{C}$: By dominated convergence, $\frac{1}{T(k)}$ is continuous in $\mathbb{C}$. (Use the bound on m_1 from Theorem 3.2). Analyticity then follows with Morera's theorem. Analyticity of α, γ and δ follows in the same way. That R_1, R_2 and T are meromorphic in $\mathbb{C}$ is now immediate from their definitions.

Finally, $\frac{1}{T(k)}$ has no zeros in $\overline{\mathbb{C}^+} = \{\operatorname{Im} z \geq 0\}$. Indeed, because of (iii) of (b), $\frac{1}{T(k)} \geq 1$ for $k \in \mathbb{R}$. Assume to the contrary that $\frac{1}{T(k)}$ has a zero at $k_0 \in \mathbb{C}^+ = \{\operatorname{Im} z > 0\}$. Then by (3.12) and by the initial conditions for u_1 and u_2, $u_1(\cdot, k_0)$ and $u_1'(\cdot, k_0)$ decay exponentially in both directions and especially lie in $L^2(\mathbb{R})$. Since u_1 solves the Helmholtz equation (3.1),

$$\begin{aligned} 0 &= \int_{-\infty}^{\infty} \left(u_1''(x,k_0) + k_0^2 n^2(x) u_1(x,k_0)\right) \overline{u_1(x,k_0)} \, \mathrm{d}x \\ &= -\int_{-\infty}^{\infty} |u_1'(x,k_0)|^2 \, \mathrm{d}x + k_0^2 \int_{-\infty}^{\infty} n^2(x) |u_1(x,k_0)|^2 \, \mathrm{d}x. \end{aligned}$$

This implies $k_0^2 > 0$, and hence $k_0 \in \mathbb{R}$, which is a contradiction. Thus, $\frac{1}{T(k)}$ has no zeros in $\overline{\mathbb{C}^+}$, whence $T(k)$ is analytic there. Writing $R_j(k) = T(k) \frac{R_j(k)}{T(k)}$, $j = 1, 2$, we see that R_1 and R_2 are also analytic on $\overline{\mathbb{C}^+}$. □

3.1.4 Further estimates

In this subsection we derive bounds on the *derivative* of the reflection and transmission coefficient on the real line. These estimates do not only guarantee that R_1, R_2 and T do not behave too wildly and thus justify that in practice these quantities are only evaluated at a finite number of points, but they are also an ingredient that is needed when proving solvability of an optimization problem in Section 3.4.

We write

$$N_1 = \int_0^d |1 - n^2(t)|\,\mathrm{d}t, \qquad N_\infty = \sup_{t\in[0,d]} |1 - n^2(t)|.$$

Theorem 3.6. *The following bounds hold ($k \in \mathbb{R}$).*

(a) *For the transmission coefficient T*

$$|T'(k)| \leq \frac{1}{2} N_1 \left(\mathrm{e}^{|k|N_1} + \left(|k|d + \frac{1}{2}|k|^2 d^2 \right) N_\infty \mathrm{e}^{2|k|N_1} \right).$$

(b) *For the reflection coefficients R_j, $j = 1, 2$,*

$$|R_j'(k)| \leq \frac{1}{2} N_1 \Big((1 + 2|k|d + |k||T'(k)|)\mathrm{e}^{|k|N_1} + \left(|k|d + (1/2)|k|^2 d^2\right) N_\infty \mathrm{e}^{2|k|N_1} \Big).$$

The exact form of the above bounds is not important. The main point is that for k in bounded intervals T' and R' can be bounded independently of k and $n \in L^\infty_{a,b}(0, d)$.

Proof. We are several times going to use the bounds

$$\sup_{x\in[0,d]} |m_j(x, k)| \leq \mathrm{e}^{|k|N_1}, \qquad j = 1, 2, \tag{3.25}$$

$$\sup_{x\in[0,d]} |\dot{m}_j(x, k)| \leq \left(d + \frac{1}{2}|k|d^2 \right) N_\infty \mathrm{e}^{2|k|N_1}, \qquad j = 1, 2, \tag{3.26}$$

which follow from Theorem 3.2.

By Theorem 3.5(c),

$$T(k) = \frac{1}{1 - F(k)},$$

where

$$F(k) = \frac{1}{2i} k \int_0^d (1 - n^2(t)) m_1(t, k) \, \mathrm{d}t.$$

Hence,

$$T'(k) = \frac{F'(k)}{(1 - F(k))^2} = T(k)^2 F'(k).$$

By Theorem 3.5, $|T(k)| \leq 1$ for $k \in \mathbb{R}$, so the above equation implies $|T'(k)| \leq |F'(k)|$. Now, using (3.25) and (3.26),

$$\begin{aligned}
|F'(k)| &\leq \frac{1}{2} \int_0^d |1 - n^2(t)| \left(|m_1(t, k)| + |k| |\dot{m}_1(t, k)| \right) \mathrm{d}t \\
&\leq \frac{1}{2} N_1 \sup_{t \in [0, d]} \left(|m_1(t, k)| + |k| |\dot{m}_1(t, k)| \right) \\
&\leq \frac{1}{2} N_1 \left(\mathrm{e}^{|k| N_1} + \left(|k| d + \frac{1}{2} |k|^2 d^2 \right) N_\infty \mathrm{e}^{2|k| N_1} \right).
\end{aligned}$$

This yields (a).

For the second estimate we use the integral representation from Theorem 3.5(c),

$$R_j(k) = T(k) G_j(k), \qquad j = 1, 2,$$

where

$$G_j(k) = \frac{1}{2i} k \int_0^d \mathrm{e}^{-2ikt} (1 - n^2(t)) m_{\tilde{j}}(t, k) \, \mathrm{d}t$$

and $\tilde{j} = 1$ for $j = 2$ and $\tilde{j} = 2$ for $j = 1$. We then have to estimate

$$R_j'(k) = T'(k) G_j(k) + T(k) G_j'(k).$$

By (3.25),

$$|G_j(k)| \leq \frac{1}{2} |k| \int_0^d |1 - n^2(t)| |m_{\tilde{j}}(t, k)| \, \mathrm{d}t \leq \frac{1}{2} |k| N_1 \mathrm{e}^{|k| N_1}.$$

Furthermore, by (3.25) and (3.26),

$$\begin{aligned}
|G_j'(k)| &\leq \frac{1}{2} \int_0^d |1 - n^2(t)| \left(|m_{\tilde{j}}(t, k)| + |k| \left(2t |m_{\tilde{j}}(t, k)| + |\dot{m}_{\tilde{j}}(t, k)| \right) \right) \mathrm{d}t \\
&\leq \frac{1}{2} N_1 \sup_{t \in [0, d]} \left((1 + 2|k| d) |m_{\tilde{j}}(t, k)| + |k| |\dot{m}_{\tilde{j}}(t, k)| \right) \\
&\leq \frac{1}{2} N_1 \left((1 + 2|k| d) \mathrm{e}^{|k| N_1} + \left(|k| d + \frac{1}{2} |k|^2 d^2 \right) N_\infty \mathrm{e}^{2|k| N_1} \right).
\end{aligned}$$

Together,

$$|R_j'(k)| \le \frac{1}{2} N_1 \Bigg((1 + 2|k|d + |k||T'(k)|) e^{|k|N_1} + \left(|k|d + (1/2)|k|^2 d^2\right) N_\infty e^{2|k|N_1} \Bigg). \qquad \square$$

3.2 Continuity of the direct problem

The aim of this section is to show that the reflection and transmission coefficients satisfy a continuity property with respect to the permittivity $\epsilon = n^2$ (Corollary 3.12). Before we can prove this property, we need to derive a somewhat more direct connection between these quantities.

3.2.1 Definition of R and T via an initial value problem

For practical purposes we are going to make the definition of reflection and transmission coefficients via (3.12) and (3.13) a little more explicit. We only consider R_2 and T, i.e., the reflection and transmission coefficient from the left.

Evaluating (3.13) at $x = 0$ and using the initial conditions (3.3),

$$u_1(0,k) = \gamma(k) + \delta(k), \tag{3.27}$$

$$u_1'(0,k) = -ik\gamma(k) + ik\delta(k). \tag{3.28}$$

We can solve this for γ and δ to get ($k \neq 0$)

$$\gamma(k) = \frac{1}{2}\left(u_1(0,k) + \frac{1}{k}u_1'(0,k)i\right),$$

$$\delta(k) = \frac{1}{2}\left(u_1(0,k) - \frac{1}{k}u_1'(0,k)i\right).$$

For reflection and transmission coefficient we thus get

$$R_2(k) = \frac{\gamma(k)}{\delta(k)} = \frac{u_1(0,k) + \frac{1}{k}u_1'(0,k)i}{u_1(0,k) - \frac{1}{k}u_1'(0,k)i}, \tag{3.29}$$

$$T(k) = \frac{1}{\delta(k)} = \frac{2}{u_1(0,k) - \frac{1}{k}u_1'(0,k)i}. \tag{3.30}$$

The values $u_1(0,k)$ and $u_1'(0,k)$ can obtained for each k by solving the Helmholtz equation (3.1) with initial conditions (3.2). In practice, one computes the reflection and transmission coefficient by solving the initial value problem numerically. If n is a step function, then the solution of the IVP boils down to the multiplication of certain matrices. The method is therefore called *transfer matrix method* [24].

3.2.2 Definition of R and T via a boundary value problem

In the above formulas, the connection between R and T on the one hand and the refractive profile n on the other hand is still somewhat indirect, since we have to first solve an initial value problem and then use (3.29) or (3.30). In the following we are going to derive a slightly more direct relationship.

Let $k \in \mathbb{C}$ such that T does not have a pole there. Define

$$u(x,k) = T(k)u_1(x,k).$$

Obviously, u solves the Helmholtz equation (3.1). Using the initial conditions (3.2) for u_1, we see that

$$iku(d,k) - u'(d,k) = 0.$$

Using (3.27) and (3.28), we have

$$iku(0,k) + u'(0,k) = ikT(k)u_1(0,k) + T(k)u_1'(0,k) = 2ikT(k)\delta(k) = 2ik.$$

Thus, u is a solution of the boundary value problem

$$\begin{cases} u''(x,k) + k^2n^2(x)u(x,k) = 0, \\ iku(d,k) - u'(d,k) = 0, \\ iku(0,k) + u'(0,k) = 2ik. \end{cases} \tag{3.31}$$

Proposition 3.7. *Fix $n \in L^\infty_{a,b}(0,d)$ and let $k \in \mathbb{C} \setminus \{0\}$ such that the transmission coefficient T corresponding to n does not have a pole there. Then the boundary value problem* (3.31) *has the unique solution $u(x,k) = T(k)u_1(x,k)$.*

Proof. Write $L_1(k)u = iku(d) - u'(d)$ and $L_2(k)u = iku(0) + u'(0)$. By (3.11), $u_1(\cdot,k)$ and $u_1(\cdot,-k)$ form a fundamental system for the Helmholtz equation. Now (3.31) is uniquely solvable if and only if (see, e.g, [65, §26.III])

$$W(k) := \begin{vmatrix} L_1(k)u_1(\cdot,k) & L_1(k)u_1(\cdot,-k) \\ L_2(k)u_1(\cdot,k) & L_2(k)u_1(\cdot,-k) \end{vmatrix} \neq 0.$$

But

$$\begin{aligned} W(k) &= \underbrace{(iku_1(d,k) - u_1'(d,k))}_{=0}(iku_1(0,-k) + u_1'(0,-k)) \\ &\qquad - (iku_1(d,-k) - u_1'(d,-k))(iku_1(0,k) + u_1'(0,k)) \\ &\overset{(3.13)}{=} -\big(iku_1(d,-k) - u_1'(d,-k)\big) \\ &\qquad \cdot \big(ik(\gamma(k)u_2(0,k) + \delta(k)u_2(0,-k)) \\ &\qquad\qquad + (\gamma(k)u_2'(0,k) + \delta(k)u_2'(0,-k))\big) \\ &= -2ike^{-ikd}(ik(\gamma(k) + \delta(k)) + (-ik\gamma(k) + ik\delta(k))) \\ &= 4k^2e^{-ikd}\delta(k) \neq 0, \end{aligned}$$

since $T = \frac{1}{\delta}$ does not have a pole at k. The theorem follows since we have already shown that $u(x,k) = T(k)u_1(x,k)$ is a solution. □

Now let u be the solution of (3.31). We then have

$$\begin{aligned} u(0,k) = T(k)u_1(0,k) &\overset{(3.15)}{=} T(k)(R_2(k)u_2(0,k) + u_2(0,-k)) \\ &= T(k)(R_2(k) + 1), \end{aligned}$$

and

$$u(d,k) = T(k)u_1(d,k) = T(k)e^{ikd},$$

so reflection and transmission coefficient are obtained from u by

$$R_2(k) = e^{-ikd}u(d,k)(u(0,k) - 1), \qquad T(k) = e^{-ikd}u(d,k). \tag{3.32}$$

3.2.3 A weak formulation

Although it is not so common for ordinary differential equations, we will derive a weak formulation of problem (3.31). Let u be a solution of (3.31). For simplicity, in the following we drop the dependence of u on k in our notation. Multiplying by $\varphi \in C^\infty(0,d)$,

$$\begin{aligned} 0 &= \int_0^d u''(x)\varphi(x) + k^2n^2(x)u(x)\varphi(x)\,\mathrm{d}x \\ &= u'(d)\varphi(d) - u'(0)\varphi(0) + \int_0^d -u'(x)\varphi'(x) + k^2n^2(x)u(x)\varphi(x)\,\mathrm{d}x \\ &= ik\left(u(d)\varphi(d) + u(0)\varphi(0)\right) \\ &\quad + \int_0^d k^2n^2(x)u(x)\varphi(x) - u'(x)\varphi'(x)\,\mathrm{d}x - 2ik\varphi(0). \end{aligned}$$

The weak form of (3.31) is then

$$\text{Find } u \in H^1(0,d) \text{ with } B_n[u,\varphi] = 2ik\varphi(0) \text{ for all } \varphi \in C^\infty(0,d), \tag{3.33}$$

where

$$B_n[u,\varphi] = ik\left(u(d)\varphi(d) + u(0)\varphi(0)\right) + \int_0^d k^2n^2(x)u(x)\varphi(x) - u'(x)\varphi'(x)\,\mathrm{d}x.$$

The careful reader should not confuse the Sobolev space $H^1(0,d)$ with the Hardy spaces $H^p(\mathbb{D})$ and $H^p(\mathbb{C}^+)$.

Theorem 3.8. *Assume that $u \in H^1(0,d)$ is a solution of (3.33). Then u has higher regularity, $u \in W^{2,\infty}(0,d)$, and u solves the classical boundary value problem (3.31).*

Proof. Let u be a solution of problem (3.33). Then $\widetilde{u}(x) = u(x) - u(0) - \frac{1}{d}(u(d) - u(0))x$ is a weak solution of

$$\begin{cases} \widetilde{u}''(x) + k^2n^2(x)\widetilde{u}(x) = \widetilde{f}(x), \\ \widetilde{u}(0) = \widetilde{u}(d) = 0, \end{cases}$$

with $\widetilde{f}(x) = -k^2n^2(x)\left(u(0) + \frac{1}{d}(u(d) - u(0))x\right)$. By standard regularity theory (see, e.g., [20, Chapter 6.3, Theorem 4]), $\widetilde{u} \in H^2(0,d)$. Thus, $u \in H^2(0,d)$, and as usual one can show that u actually solves the classical boundary value problem (3.31). Finally, since $u''(x) = -k^2n^2(x)u(x)$, we even have $u \in W^{2,\infty}(0,d)$. □

Remark 3.9. *By Theorem 3.8, any weak solution is also a classical solution. Since we have seen that conversely any classical solution is a weak solution, problem (3.31) and its weak formulation (3.33) are equivalent. Especially, we have unique solvability of the weak problem if $k \in \mathbb{C} \setminus \{0\}$ is not a pole of the transmission coefficient.*

3.2.4 Continuity in the weak* topology of L^∞

After these preparations we can prove the promised continuity property. We are not going to work with the refractive index n, but with the permittivity $\epsilon = n^2$. Denote by S_k the mapping which sends $\epsilon \in L^\infty_{a,b}(0,d)$ to the solution $S_k\epsilon$ of the boundary value problem (3.33) (or, equivalently, (3.31)) with $n = \sqrt{\epsilon}$ and k, i.e.,

$$B_{\sqrt{\epsilon}}[S_k\epsilon, \varphi] = 2ik\varphi(0) \text{ for all } \varphi \in C^\infty(0,d).$$

We first show continuity of S_k in the weak* topology of L^∞. It is interesting to compare the following theorem to [19, Theorem 3.1], which claims weak* continuity of the solution of a similar boundary value problem with respect to n. However, this is not correct, but weak* continuity with respect to $\epsilon = n^2$ holds true. (For a counterexample see Remark 3.13 below.)

Theorem 3.10. *Fix $k \in \mathbb{R} \setminus \{0\}$. The mapping $S_k : L^\infty_{a,b}(0,d) \to H^1(0,d)$, $\epsilon \mapsto S_k\epsilon$, where $S_k\epsilon$ is the solution of (3.33) with $n = \sqrt{\epsilon}$, is (sequentially) continuous in the weak* topology of L^∞ and the weak topology of $H^1(0,d)$, i.e., $(\epsilon_j) \subset L^\infty_{a,b}(0,d)$ with $\epsilon_j \overset{*}{\rightharpoonup} \epsilon$ implies $S_k\epsilon_j \rightharpoonup S_k\epsilon$ in $H^1(0,d)$.*

Proof. Let $(\epsilon_j) \subset L^\infty_{a,b}(0,d)$ with $\epsilon_j \overset{*}{\rightharpoonup} \epsilon$, i.e.,

$$\int_0^d \epsilon_j(x)\varphi(x)\,\mathrm{d}x \to \int_0^d \epsilon(x)\varphi(x)\,\mathrm{d}x$$

for all $\varphi \in L^1(0,d)$.

Step 1 (Find weakly converging subsequence of $(S_k\epsilon_j)$): By Theorem 3.5(a) the transmission coefficient has no pole in k. So by Proposition 3.7, the solution of (3.33) is given by $(S_k\epsilon)(x) = T(k)u_1(x,k)$. We are going to show that $\|S_k\epsilon\|_{L^\infty(0,d)}$ and $\|(S_k\epsilon)'\|_{L^\infty(0,d)}$ can be bounded by a constant independent of $\epsilon \in L^\infty_{a,b}(0,d)$.

Since $k \in \mathbb{R}$, we have $|T(k)| \leq 1$ by Theorem 3.5(b)(iii). By Theorem 3.2(a) we have for $x \in [0,d]$

$$|u_1(x,k)| = |\mathrm{e}^{ikx}m_1(x,k)| = |m_1(x,k)| \leq \mathrm{e}^{|k|\int_x^d |1-n^2(t)|\,\mathrm{d}t} \leq \mathrm{e}^{|k|d(b+1)},$$

so $\|S_k\epsilon\|_{L^\infty(0,d)} \leq C_1$ with $C_1 = \mathrm{e}^{|k|d(b+1)}$ for all $\epsilon \in L^\infty_{a,b}(0,d)$. Further, from

(3.4) one obtains by differentiation

$$\begin{aligned} m_1'(x,k) &= \int_x^d m_1(t,k)(1-n^2(t))(-k^2 \mathrm{e}^{2ik(t-x)})\,\mathrm{d}t \\ &\qquad - m_1(x,k)(1-n^2(x))\underbrace{D_k(0)}_{=0} \\ &= \int_x^d m_1(t,k)(1-n^2(t))(-k^2 \mathrm{e}^{2ik(t-x)})\,\mathrm{d}t. \end{aligned}$$

So for $x \in [0,d]$

$$|m_1'(x,k)| \le dC_1(b+1)|k|^2,$$

that is, $\|(S_k\epsilon)'\|_{L^\infty(0,d)} \le C_2$ with $C_2 = dC_1(b+1)|k|^2$ for all $\epsilon \in L^\infty_{a,b}(0,d)$. This means that the sequence $(S_k\epsilon_j)$ is bounded in $W^{1,\infty}(0,d)$.

Especially, $(S_k\epsilon_j)$ is bounded in $H^1(0,d)$, so it has a weakly convergent subsequence $(S_k\epsilon_{j_l})$, i.e., $S_k\epsilon_{j_l} \rightharpoonup u$ in $H^1(0,d)$ for some $u \in H^1(0,d)$.

Step 2 (Show that $S_k\epsilon = u$): We show that $B_{\sqrt{\epsilon}}[S_k\epsilon,\varphi] = B_{\sqrt{\epsilon}}[u,\varphi]$ for all $\varphi \in C^\infty(0,d)$. Unique solvability of (3.33) then implies $S_k\epsilon = u$. First notice that $B_{\sqrt{\epsilon}}[S_k\epsilon,\varphi] = 2ik\varphi(0) = B_{\sqrt{\epsilon_{j_l}}}[S_k\epsilon_{j_l},\varphi]$ for each j_l, so

$$\begin{aligned} &\left|B_{\sqrt{\epsilon}}[u,\varphi] - B_{\sqrt{\epsilon}}[S_k\epsilon,\varphi]\right| \\ &\quad = \left|B_{\sqrt{\epsilon}}[u,\varphi] - B_{\sqrt{\epsilon_{j_l}}}[S_k\epsilon_{j_l},\varphi]\right| \\ &\quad \le \left|B_{\sqrt{\epsilon}}[u,\varphi] - B_{\sqrt{\epsilon_{j_l}}}[u,\varphi]\right| + \left|B_{\sqrt{\epsilon_{j_l}}}[u,\varphi] - B_{\sqrt{\epsilon_{j_l}}}[S_k\epsilon_{j_l},\varphi]\right|. \end{aligned}$$

For the first term we have

$$\left|B_{\sqrt{\epsilon}}[u,\varphi] - B_{\sqrt{\epsilon_{j_l}}}[u,\varphi]\right| = \left|\int_0^d k^2\left(\epsilon(x) - \epsilon_{j_l}(x)\right)u(x)\varphi(x)\,\mathrm{d}x\right| \to 0,$$

since $u\varphi \in L^1(0,d)$. For the second term we have

$$\begin{aligned} &\left|B_{\sqrt{\epsilon_{j_l}}}[u,\varphi] - B_{\sqrt{\epsilon_{j_l}}}[S_k\epsilon_{j_l},\varphi]\right| \\ &\quad = \left|ik\left((S_k\epsilon_{j_l} - u)(d)\varphi(d) + (S_k\epsilon_{j_l} - u)(0)\varphi(0)\right)\right| \\ &\qquad + \left|\int_0^d k^2\epsilon_{j_l}(x)(S_k\epsilon_{j_l} - u)(x)\varphi(x) - ((S_k\epsilon_{j_l})' - u')(x)\varphi'(x)\,\mathrm{d}x\right| \\ &\quad \to 0. \end{aligned}$$

Here, the second summand converges to zero since $S_k\epsilon_{j_l} \rightharpoonup u$ in $H^1(0,d)$, and the first summand converges to zero since especially $S_k\epsilon_{j_l} \to u$ in $C(0,d)$.

(This follows from the compact embedding $H^1(0,d) \hookrightarrow C(0,d)$.) Together, it follows that $\left|B_{\sqrt{\epsilon}}[u,\varphi] - B_{\sqrt{\epsilon}}[S_k\epsilon,\varphi]\right| = 0$, which is what we wanted to prove in this step.

Step 3 (Show $S_k\epsilon_j \rightharpoonup S_k\epsilon$): It now follows that the whole sequence $S_k\epsilon_j$ converges weakly to u in $H^1(0,d)$: Indeed, if there were infinitely many $S_k\epsilon_j$ outside of an arbitrary (weak $H^1(0,d)$-)neighborhood of u, we could apply the preceding arguments to find a subsequence of these infinitely many $S_k\epsilon_j$ which converges to u, thus producing a contradiction. This proves the theorem. □

Remark 3.11. *The reason why we could only prove the theorem for real k is that we only have the bound $|T(k)| \leq 1$ for real k. We will use the same technique as in this proof to prove Theorem 3.19, but there we will have a bound for $|T(k)|$, $k \in \mathbb{C}^+$.*

The following corollary is the culmination of this section and a direct consequence of the preceding theorem and (3.32).

Corollary 3.12. *At each $k \in \mathbb{R}$, reflection and transmission coefficient are (sequentially) weak* continuous functions of ϵ, i.e., if $R_{\sqrt{\epsilon}}$ and $T_{\sqrt{\epsilon}}$ denote the reflection and transmission coefficient (from the left) corresponding to the permittivity ϵ, then $(\epsilon_j) \subset L^\infty_{a,b}(0,d)$ with $\epsilon_j \overset{*}{\rightharpoonup} \epsilon$ implies $R_{\sqrt{\epsilon_j}}(k) \to R_{\sqrt{\epsilon}}(k)$ and $T_{\sqrt{\epsilon_j}}(k) \to T_\epsilon(k)$.*

Proof. Let $(\epsilon_j) \subset L^\infty_{a,b}(0,d)$ with $\epsilon_j \overset{*}{\rightharpoonup} \epsilon$. If $k = 0$, then $R_{\sqrt{\epsilon_j}}(k) = R_{\sqrt{\epsilon}}(k) = 1$ and $T_{\sqrt{\epsilon_j}}(k) = T_{\sqrt{\epsilon}}(k) = 0$, so there is nothing to prove. Assume $k \in \mathbb{R} \setminus \{0\}$. From Theorem 3.10 it follows that S_k is especially (sequentially) weak* continuous as a map to $C(0,d)$, i.e., $\epsilon_j \overset{*}{\rightharpoonup} \epsilon$ in $L^\infty_{a,b}(0,d)$ implies $S_k\epsilon_j \to S_k\epsilon$ in $C(0,d)$. This is due to the fact that the embedding $H^1(0,d) \hookrightarrow C(0,d)$ is compact and that weakly convergent sequences become strongly convergent under compact mappings. From (3.32) we then get

$$T_{\sqrt{\epsilon_j}}(k) = \mathrm{e}^{-ikd}(S_k\epsilon_j)(d) \to \mathrm{e}^{-ikd}(S_k\epsilon)(d) = T_{\sqrt{\epsilon}}(k)$$

and

$$R_{\sqrt{\epsilon_j}}(k) = \mathrm{e}^{-ikd}T_{\epsilon_j}(k)((S_k\epsilon_j)(0)-1) \to \mathrm{e}^{-ikd}T_\epsilon(k)((S_k\epsilon)(0)-1) = R_{\sqrt{\epsilon}}(k).$$ □

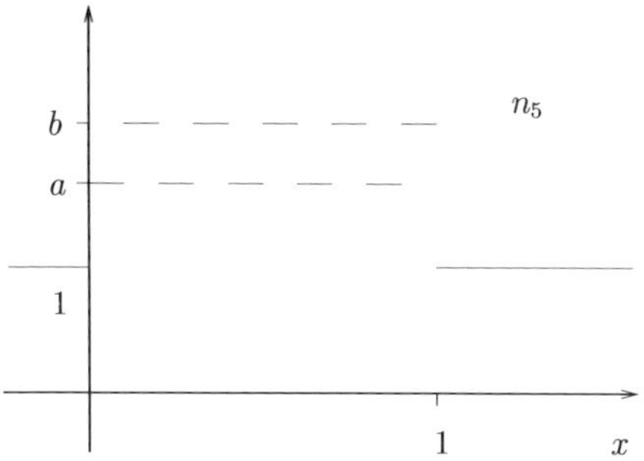

Figure 3.2: The function n_j, $j = 5$, from Remark 3.13.

Remark 3.13. *The convergence of Corollary 3.12 can of course also be observed numerically. For example, let us set for some $a, b > 0$*

$$n_j(x) = \begin{cases} a + (b-a)\left(\lfloor 2jx \rfloor \bmod 2\right), & x \in [0,1], \\ 1, & \text{otherwise}, \end{cases}$$

see Figure 3.2. Then $n_j^2 \overset{*}{\rightharpoonup} n^2$, where $n^2(x) = (a^2 + b^2)/2$, $x \in [0,1]$, and $n^2(x) = 1$, otherwise. Indeed, if one does the computations, one sees that $R_{n_j}(k) \to R_n(k)$, and not, as one might naively expect, $R_{n_j}(k) \to R_{\widetilde{n}}(k)$, where $\widetilde{n}(x) = (a+b)/2$, $x \in [0,1]$, and $\widetilde{n}(x) = 1$, otherwise.

3.3 Hardy space properties

In this section we show that the reflection coefficient lies in the Hardy space $H^\infty(\mathbb{C}^+)$. Although this appears reasonable, we are not aware of any proof of our results in the literature. We already know that the reflection coefficient is analytic on $\mathbb{C}^+$ (Theorem 3.5(a)), but it remains to show that it is bounded. We first show this property when the refractive profile n is a step function and then use the method from Section 3.2.4 to extend the result to general n. We begin with some preparations.

3.3.1 Changing the surrounding medium

In this and the next subsection we investigate how reflection and and transmission coefficient vary under certain transformations of n. Assume that a refractive profile n is given with $n|_{[0,d]} \in L^\infty_{a,b}(0,d)$, but now $n|_{(-\infty,0)} = n_l$ and $n|_{(d,\infty)} = n_r$, where n_l and n_r are positive, but not necessarily equal to 1. It is straightforward to generalize the definitions of reflection and transmission coefficient to this case: Let $u_{1,n_r}(x,k)$ be the solution of the Helmholtz equation (3.1) with initial conditions

$$\begin{cases} u_{1,n_r}(d,k) = \mathrm{e}^{ikn_r d}, \\ u'_{1,n_r}(d,k) = ikn_r \mathrm{e}^{ikn_r d}, \end{cases} \tag{3.34}$$

and let $u_{2,n_l}(x,k)$ be the solution of (3.1) with initial conditions

$$\begin{cases} u_{2,n_l}(0,k) = 1, \\ u'_{2,n_l}(0,k) = -ikn_l. \end{cases} \tag{3.35}$$

We now have $u_{1,n_r}(x,k) = \mathrm{e}^{ikn_r x}$ for $x \geq d$ and $u_{2,n_l}(x,k) = \mathrm{e}^{-ikn_l x}$ for $x \leq 0$. One can see as in Section 3.1.3 that $u_j(\cdot,k)$ and $u_j(\cdot,-k)$ are linearly independent for $k \neq 0$, $j = 1,2$, whence there are functions $\alpha_{n_l,n_r}(k)$, $\beta_{n_l,n_r}(k)$, $\gamma_{n_l,n_r}(k)$ and $\delta_{n_l,n_r}(k)$ such that

$$u_{2,n_l}(x,k) = \alpha_{n_l,n_r}(k) u_{1,n_r}(x,k) + \beta_{n_l,n_r}(k) u_{1,n_r}(x,-k), \tag{3.36}$$
$$u_{1,n_r}(x,k) = \gamma_{n_l,n_r}(k) u_{2,n_l}(x,k) + \delta_{n_l,n_r}(k) u_{2,n_l}(x,-k) \tag{3.37}$$

for $k \neq 0$. As earlier, reflection and transmission coefficient from the left and from the right are defined by

$$T_{1,n_l,n_r}(k) = \frac{1}{\beta_{n_l,n_r}(k)}, \qquad R_{1,n_l,n_r}(k) = \frac{\alpha_{n_l,n_r}(k)}{\beta_{n_l,n_r}(k)},$$
$$T_{2,n_l,n_r}(k) = \frac{1}{\delta_{n_l,n_r}(k)}, \qquad R_{2,n_l,n_r}(k) = \frac{\gamma_{n_l,n_r}(k)}{\delta_{n_l,n_r}(k)}.$$

The following proposition states how reflection and transmission coefficient from the left change if we change the surrounding medium on the left, i.e., n_l.

Proposition 3.14. *Let n_l, $\widetilde{n}_l > 0$. For transmission and reflection coefficient from the left with respect to n_l and $\widetilde{n}_l$, respectively, we have*

$$R_{2,\widetilde{n}_l,n_r}(k) = \frac{\left(1 + \frac{n_l}{\widetilde{n}_l}\right) R_{2,n_l,n_r}(k) + \left(1 - \frac{n_l}{\widetilde{n}_l}\right)}{\left(1 - \frac{n_l}{\widetilde{n}_l}\right) R_{2,n_l,n_r}(k) + \left(1 + \frac{n_l}{\widetilde{n}_l}\right)},$$
$$T_{2,\widetilde{n}_l,n_r}(k) = \frac{2T_{2,n_l,n_r}}{\left(1 - \frac{n_l}{\widetilde{n}_l}\right) R_{2,n_l,n_r}(k) + \left(1 + \frac{n_l}{\widetilde{n}_l}\right)}.$$

Proof. As in Section 3.2.1, evaluating (3.37) at $x = 0$ and using the initial conditions (3.35),

$$u_{1,n_r}(x,k) = \gamma_{n_l,n_r}(k) + \delta_{n_l,n_r}(k), \tag{3.38}$$

$$u'_{1,n_r}(0,k) = -ikn_l\gamma_{n_l,n_r}(k) + ikn_r\delta_{n_l,n_r}(k). \tag{3.39}$$

Notice that the left hand side does not depend on n_l. We solve this for γ_{n_l,n_r} and δ_{n_l,n_r} to get ($k \neq 0$)

$$\gamma_{n_l,n_r}(k) = \frac{1}{2}\left(u_{1,n_r}(0,k) + \frac{1}{kn_l}u'_{1,n_r}(0,k)i\right), \tag{3.40}$$

$$\delta_{n_l,n_r}(k) = \frac{1}{2}\left(u_{1,n_r}(0,k) - \frac{1}{kn_l}u'_{1,n_r}(0,k)i\right). \tag{3.41}$$

We now plug (3.38) and (3.39) into (3.40) and (3.41) for $\widetilde{n}_l$ instead of n_l and obtain

$$\gamma_{\widetilde{n}_l,n_r}(k) = \frac{1}{2}\left(\left(1 + \frac{n_l}{\widetilde{n}_l}\right)\gamma_{n_l,n_r}(k) + \left(1 - \frac{n_l}{\widetilde{n}_l}\right)\delta_{n_l,n_r}(k)\right),$$

$$\delta_{\widetilde{n}_l,n_r}(k) = \frac{1}{2}\left(\left(1 - \frac{n_l}{\widetilde{n}_l}\right)\gamma_{n_l,n_r}(k) + \left(1 + \frac{n_l}{\widetilde{n}_l}\right)\delta_{n_l,n_r}(k)\right).$$

It follows that

$$\begin{aligned} R_{2,\widetilde{n}_l,n_r}(k) = \frac{\gamma_{\widetilde{n}_l,n_r}(k)}{\delta_{\widetilde{n}_l,n_r}(k)} &= \frac{\left(1 + \frac{n_l}{\widetilde{n}_l}\right)\gamma_{n_l,n_r}(k) + \left(1 - \frac{n_l}{\widetilde{n}_l}\right)\delta_{n_l,n_r}(k)}{\left(1 - \frac{n_l}{\widetilde{n}_l}\right)\gamma_{n_l,n_r}(k) + \left(1 + \frac{n_l}{\widetilde{n}_l}\right)\delta_{n_l,n_r}(k)} \\ &= \frac{\left(1 + \frac{n_l}{\widetilde{n}_l}\right)R_{2,n_l,n_r}(k) + \left(1 - \frac{n_l}{\widetilde{n}_l}\right)}{\left(1 - \frac{n_l}{\widetilde{n}_l}\right)R_{2,n_l,n_r}(k) + \left(1 + \frac{n_l}{\widetilde{n}_l}\right)} \end{aligned}$$

and

$$\begin{aligned} T_{2,\widetilde{n}_l,n_r}(k) = \frac{1}{\delta_{\widetilde{n}_l,n_r}(k)} &= \frac{2}{\left(1 - \frac{n_l}{\widetilde{n}_l}\right)\gamma_{n_l,n_r}(k) + \left(1 + \frac{n_l}{\widetilde{n}_l}\right)\delta_{n_l,n_r}(k)} \\ &= \frac{2T_{2,n_l,n_r}(k)}{\left(1 - \frac{n_l}{\widetilde{n}_l}\right)R_{2,n_l,n_r}(k) + \left(1 + \frac{n_l}{\widetilde{n}_l}\right)}. \end{aligned}$$

□

It is straightforward to modify the proof of Theorem 3.5(b) to obtain

Proposition 3.15. *We have the following relations:*

(a) *For* $k \in \mathbb{C}$, $n_l T_{1,n_l,n_r}(k) = n_r T_{2,n_l,n_r}(k)$.

(b) *For* $k \in \mathbb{C}$, *the reality conditions*

$$\overline{T_{j,n_l,n_r}(k)} = T_{j,n_l,n_r}(-\overline{k})$$

and

$$\overline{R_{j,n_l,n_r}(k)} = R_{j,n_l,n_r}(-\overline{k}),$$

$j = 1, 2$, *hold.*

(c) *For* $k \in \mathbb{R}$,

$$\frac{n_l}{n_r}|T_{1,n_l,n_r}(k)|^2 + |R_{1,n_l,n_r}(k)|^2 = \frac{n_r}{n_l}|T_{2,n_l,n_r}(k)|^2 + |R_{2,n_l,n_r}(k)|^2 = 1.$$

Notice that T_1 and T_2 do not coincide any more if $n_l \neq n_r$.

3.3.2 Shifting n

We again consider a refractive profile n with $n|_{[0,d]} \in L^{\infty}_{a,b}(0,d)$ and $n|_{(-\infty,0)} = n_l$ and $n|_{(d,\infty)} = n_r$, where n_l and n_r are positive real numbers. We are interested in reflection and transmission coefficient of the shifted profile $n^s(x) = n(x - d_1)$, where $d_1 > 0$. In this subsection, all quantities with respect to n^s are decorated with a small s. Especially, R_2^s and T_2^s denote reflection and transmission coefficient from the left with respect to n^s, while R_2 and T_2 denote reflection and transmission coefficient from the left with respect to n. (For simplicity, we drop the dependence on n_l and n_r in our notation.)

Proposition 3.16. *We have*

$$R_2^s(k) = \mathrm{e}^{2ikn_l d_1} R_2(k),$$
$$T_2^s(k) = \mathrm{e}^{ik(n_l - n_r)d_1} T_2(k).$$

Proof. From the initial conditions (3.34) for u_1 at $x = d$ and for u_1^s at $x = d + d_1$ it is not hard to see that

$$u_1^s(x + d_1, k) = \mathrm{e}^{ikn_r d_1} u_1(x, k).$$

From (3.37) we have

$$\begin{aligned} u_1(x,k) &= \gamma(k)\mathrm{e}^{-ikn_l x} + \delta(k)\mathrm{e}^{ikn_l x} && \text{for } x \le 0, \\ u_1^s(x,k) &= \gamma^s(k)\mathrm{e}^{-ikn_l x} + \delta^s(k)\mathrm{e}^{ikn_l x} && \text{for } x \le d_1, \end{aligned}$$

so on the one hand

$$\begin{aligned} u_1^s(0,k) &= \gamma^s(k) + \delta^s(k), \\ (u_1^s)'(0,k) &= -ikn_l\gamma^s(k) + ikn_l\delta^s(k), \end{aligned}$$

and on the other hand

$$\begin{aligned} u_1^s(0,k) &= \mathrm{e}^{ikn_r d_1} u_1(-d_1,k) = \mathrm{e}^{ik(n_r+n_l)d_1}\gamma(k) + \mathrm{e}^{ik(n_r-n_l)d_1}\delta(k), \\ (u_1^s)'(0,k) &= \mathrm{e}^{ikn_r d_1} u_1'(-d_1,k) = -ikn_l\mathrm{e}^{ik(n_r+n_l)d_1}\gamma(k) + ikn_l\mathrm{e}^{ik(n_r-n_l)d_1}\delta(k). \end{aligned}$$

Together,

$$\begin{aligned} \gamma^s(k) &= \mathrm{e}^{ik(n_r+n_l)d_1}\gamma(k), \\ \delta^s(k) &= \mathrm{e}^{ik(n_r-n_l)d_1}\delta(k). \end{aligned}$$

It now follows that

$$R_2^s(k) = \frac{\gamma^s(k)}{\delta^s(k)} = \mathrm{e}^{2ikn_l d_1}\frac{\gamma(k)}{\delta(k)} = \mathrm{e}^{2ikn_l d_1} R_2(k)$$

and

$$T_2^s(k) = \frac{1}{\delta^s(k)} = \frac{1}{\mathrm{e}^{ik(n_r-n_l)d_1}}\delta(k) = \mathrm{e}^{ik(n_l-n_r)d_1} T_2(k). \qquad \square$$

3.3.3 Hardy space properties of R and T

After these preparations we can finally show that the reflection and transmission coefficient lie in certain Hardy spaces. As already mentioned, analyticity follows from Theorem 3.5(a), and it remains to prove boundedness. We are going to do this in two steps. First, we show boundedness of R and T for piecewise constant refractive indices n using the results from Sections 3.3.1 and 3.3.2. We then generalize this result to arbitrary $n \in L^\infty_{a,b}(0,d)$ via density and the method of Section 3.2.4.

We need some notation. Assume that we have a refractive profile n with $n|_{[0,d]} \in L^\infty_{a,b}(0,d)$ and $n|_{\mathbb{R}\setminus[0,d]} = 1$. By adding a layer we formally mean

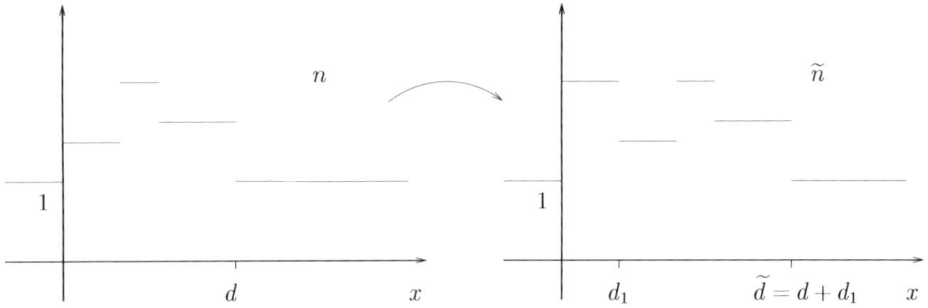

Figure 3.3: Adding a layer.

replacing n by $\widetilde{n}$, where

$$\widetilde{n}(x) = \begin{cases} 1, & x < 0, \\ n_1, & 0 \leq x \leq d_1, \\ n(x - d_1), & x > d_1, \end{cases}$$

see Figure 3.3. Here, $n_1, d_1 > 0$. We can also describe the change from n to $\widetilde{n}$ in the following way. Write

$$U_{n_l} n(x) = \begin{cases} n_l, & x < 0, \\ n(x), & x \geq 0 \end{cases}$$

for the change of the surrounding medium on the left hand side. Moreover, write

$$V_y n(x) = n(x - y)$$

for the shift as in the previous subsection. Then

$$\widetilde{n}(x) = U_1 V_{d_1} U_{n_1} n(x).$$

By R_n and T_n we denote the reflection and transmission coefficient from the left (i.e., R_2 and T_2) with respect to the refractive profile n.

Theorem 3.17. *Assume that n is a step function with $n|_{[0,d]} \in L^\infty_{a,b}(0, d)$ and $n|_{\mathbb{R}\setminus[0,d]} = 1$. Denote by $d_{\text{opt}} = \int_0^d n(x)\,\mathrm{d}x$ the optical thickness. Then*

$$R_n \in H^\infty(\mathbb{C}^+) \quad \textit{with} \quad \|R_n\|_{H^\infty(\mathbb{C}^+)} \leq 1$$

and

$$T_n \in \mathrm{e}^{i(d_{\text{opt}} - d)\cdot} H^\infty(\mathbb{C}^+) \quad \textit{with} \quad \|T_n(\cdot)\mathrm{e}^{-i(d_{\text{opt}} - d)\cdot}\|_{H^\infty(\mathbb{C}^+)} \leq 1.$$

Remark 3.18. *The impulse response of the LTI system associated with T_n is $\widehat{T_n}$. From Theorem 3.17 it follows that* $\operatorname{supp}\widehat{T_n} \subset [d_{\mathrm{opt}} - d, \infty)$. *For actual physical materials we always have $n(x) \geq 1$ for $x \in \mathbb{R}$, i.e., the optical thickness is larger than the physical thickness, $d_{\mathrm{opt}} \geq d$. In this case the LTI system associated with T_n is causal. If the optical thickness is smaller than the physical thickness (which is physically impossible), then the impulse response of the system has an acausal part.*

Proof of Theorem 3.17. We prove the theorem by induction. If $n(x) = 1$ for $x \in \mathbb{R}$, then $R_n(k) = 0$ and $T_n(k) = 1$ for all $k \in \mathbb{C}^+$. This follows for example from Theorem 3.5. Then $d = d_{\mathrm{opt}} = 0$, and it is clear that $R_n \in H^\infty(\mathbb{C}^+)$ and $T_n \in \mathrm{e}^{i(d_{\mathrm{opt}}-d)\cdot} H^\infty(\mathbb{C}^+)$ and that the norm estimates hold.

Now assume that $n \in L^\infty_{a,b}(0,d)$ is a step function. By induction hypothesis, $R_n \in H^\infty(\mathbb{C}^+)$ and $T_n \in \mathrm{e}^{i(d_{\mathrm{opt}}-d)\cdot} H^\infty(\mathbb{C}^+)$. We need to show that if we add a layer, i.e., if we replace n by $\widetilde{n}(x) = U_1 V_{d_1} U_{n_1} n(x)$, then $R_{\widetilde{n}} \in H^\infty(\mathbb{C}^+)$ and $T_{\widetilde{n}} \in \mathrm{e}^{i(\widetilde{d_{\mathrm{opt}}}-\widetilde{d})\cdot} H^\infty(\mathbb{C}^+)$, and the norm estimates still hold. Here, $\widetilde{d} = d + d_1$, and $\widetilde{d_{\mathrm{opt}}} = \int_0^{\widetilde{d}} \widetilde{n}(x)\,\mathrm{d}x = d_{\mathrm{opt}} + n_1 d_1$.

By Proposition 3.14,

$$R_{U_{n_1}n}(k) = \frac{\left(1 + \frac{1}{n_1}\right) R_n(k) + \left(1 - \frac{1}{n_1}\right)}{\left(1 - \frac{1}{n_1}\right) R_n(k) + \left(1 + \frac{1}{n_1}\right)}.$$

By induction hypothesis, $|R_n(k)| \leq 1$ on $\mathbb{C}^+$. Therefore,

$$\begin{aligned}\left|\left(1 - \frac{1}{n_1}\right) R_n(k) + \left(1 + \frac{1}{n_1}\right)\right| &\geq \left|1 + \frac{1}{n_1}\right| - \left|1 - \frac{1}{n_1}\right| \\ &= 2\min\left\{1, \frac{1}{n_1}\right\},\end{aligned} \tag{3.42}$$

whence

$$\left|R_{U_{n_1}n}(k)\right| \leq \frac{1 + \frac{1}{n_1}}{\min\left\{1, \frac{1}{n_1}\right\}}.$$

Thus, $R_{U_{n_1}n} \in H^\infty(\mathbb{C}^+)$. By Proposition 3.15, $|R_{U_{n_1}n}(k)| \leq 1$ for $k \in \mathbb{R}$. Since the modulus of functions in $H^\infty(\mathbb{C}^+)$ takes its maximum on $\mathbb{R}$,

$\|R_{U_{n_1}n}\|_{H^\infty(\mathbb{C}^+)} \leq 1$. Further, by Propositions 3.14 and 3.16,

$$R_{U_1 V_{d_1} U_{n_1} n}(k) = \frac{\left(1 + \frac{1}{n_1}\right) \mathrm{e}^{2ikn_1d_1} R_{U_{n_1}n}(k) + \left(1 - \frac{1}{n_1}\right)}{\left(1 - \frac{1}{n_1}\right) \mathrm{e}^{2ikn_1d_1} R_{U_{n_1}n}(k) + \left(1 + \frac{1}{n_1}\right)}.$$

Since $|\mathrm{e}^{2ikn_1d_1} R_{U_{n_1}n}(k)| \leq 1$ for $\operatorname{Im} k \geq 0$, it follows just as before that $\|R_{\tilde{n}}(k)\|_{H^\infty(\mathbb{C}^+)} \leq 1$.

Similarly, for the transmission coefficient we have

$$T_{U_{n_1}n}(k) = \frac{2T_n}{\left(1 - \frac{1}{n_1}\right) R_n(k) + \left(1 + \frac{1}{n_1}\right)}.$$

By (3.42), the denominator is bounded away from zero on $\mathbb{C}^+$, and by the induction hypothesis, $T_n \in \mathrm{e}^{i(d_{\mathrm{opt}}-d)\cdot} H^\infty(\mathbb{C}^+)$. Therefore, it follows that $T_{U_{n_1}n} \in \mathrm{e}^{i(d_{\mathrm{opt}}-d)\cdot} H^\infty(\mathbb{C}^+)$. Further,

$$T_{U_1 V_{d_1} U_{n_1} n}(k) = \frac{\mathrm{e}^{ik(n_1-1)d_1} T_{U_{n_1}n}(k)}{\left(1 - \frac{1}{n_1}\right) \mathrm{e}^{2ikn_1d_1} R_{U_{n_1}n}(k) + \left(1 + \frac{1}{n_1}\right)}.$$

As before, the denominator is bounded away from zero on $\mathbb{C}^+$. Using this and the fact that $T_{U_{n_1}n} \in \mathrm{e}^{i(d_{\mathrm{opt}}-d)\cdot} H^\infty(\mathbb{C}^+)$, we get

$$T_{\tilde{n}} \in \mathrm{e}^{i(d_{\mathrm{opt}}-d+(n_1-1)d_1)\cdot} H^\infty(\mathbb{C}^+) = \mathrm{e}^{i(\widetilde{d_{\mathrm{opt}}}-\tilde{d})\cdot} H^\infty(\mathbb{C}^+).$$

Moreover, since $|T_{\tilde{n}}(k)| \leq 1$ on $\mathbb{R}$ and since the modulus of functions in $H^\infty(\mathbb{C}^+)$ takes its maximum on $\mathbb{R}$, $\|T_{\tilde{n}}(\cdot)\mathrm{e}^{-i(d_{\mathrm{opt}}-d)\cdot}\|_{H^\infty(\mathbb{C}^+)} \leq 1$. This proves the induction step and therefore the theorem. □

Theorem 3.19. *Let $n|_{[0,d]} \in L^\infty_{a,b}(0,d)$ and $n|_{\mathbb{R}\setminus[0,d]} = 1$. Then we have*

$$R_n \in H^\infty(\mathbb{C}^+) \quad \textit{with} \quad \|R_n\|_{H^\infty(\mathbb{C}^+)} \leq 1$$

and

$$T_n \in \mathrm{e}^{i(a-1)d\cdot} H^\infty(\mathbb{C}^+) \quad \textit{with} \quad \|T_n(\cdot)\mathrm{e}^{-i(a-1)d\cdot}\|_{H^\infty(\mathbb{C}^+)} \leq 1.$$

Proof. Let $(n_j) \subset L^\infty_{a,b}(0,d)$ be a sequence of step functions such that for $\epsilon_j = n_j^2$ and $\epsilon = n^2$ we have $\epsilon_j \overset{*}{\rightharpoonup} \epsilon$ in $L^\infty_{a^2,b^2}(0,d)$. Such a sequence exists because step functions with values in $[a^2, b^2]$ are dense in $L^\infty_{a^2,b^2}(0,d)$ with respect to the weak* topology, see, e.g., [17, Proposition 2.2]. We then have $d_{\text{opt}}^{(j)} = \int_0^d n_j(t)\,dt \geq ad$ for all j. By Theorem 3.17, $T_{n_j} \in e^{i(d_{\text{opt}}^{(j)}-d)\cdot} H^\infty(\mathbb{C}^+)$. Because for $d_1 \geq d_2$ the inclusion $e^{id_1\cdot} H^\infty(\mathbb{C}^+) \subset e^{id_2\cdot} H^\infty(\mathbb{C}^+)$ holds true, it follows that $T_{n_j} \in e^{i(a-1)d\cdot} H^\infty(\mathbb{C}^+)$ for all j.

Now fix $k \in \mathbb{C}^+$. Then $|T_{n_j}(k)| \leq |e^{-ik(a-1)d}|$ for all j, i.e., the sequence $(T_{n_j}(k))_j$ is bounded. Let S_k be the operator from Theorem 3.10. It then follows as in Step 1 of the proof of Theorem 3.10 that $(S_k \epsilon_j)$ is bounded in $H^1(0,d)$. As in the rest of the proof of Theorem 3.10 one can show that $S_k \epsilon_j \rightharpoonup S_k \epsilon$ in $H^1(0,d)$.

Especially, it follows as in Corollary 3.12 that $R_{n_j}(k) \to R_n(k)$ and $T_{n_j}(k) \to T_n(k)$ for every $k \in \mathbb{C}^+$. By Theorem 3.17 we know that $|R_{n_j}(k)| \leq 1$, whence $|R_n(k)| \leq 1$. Similarly, $|T_n(k)| \leq |e^{-ik(a-1)d}|$. The theorem follows since we already know that R_n and T_n are analytic on $\mathbb{C}^+$. □

3.4 An optimization problem for the reflection coefficient

We have seen in the previous sections that the set of reflection coefficients which correspond to physically realizable layered media (i.e., which have a finite thickness d and a refractive profile that only varies between two bounds a and b) is severely restricted. Therefore, one can not expect that there is a refractive profile n creating an arbitrary prescribed reflection coefficient R_{desired}. As earlier, we denote by R_n the reflection coefficient from the left (i.e., R_2) corresponding to the n. Instead of using a complicated merit function like (0.1) to measure the distance between R_{desired} and a realizable reflection coefficient R_n, we simply consider the L^p-distance. To make things concrete, suppose we are given an interval $I \subset \mathbb{R}$ and a desired (complex-valued) reflection coefficient $R_{\text{desired}} \in L^\infty(I)$. Fix a thickness d and the bounds a and b. We are then interested in the minimization problem

$$\begin{array}{ll} \text{minimize} & \|R_n - R_{\text{desired}}\|_{L^p(I)} \\ \text{subject to} & n \in L^\infty_{a,b}(0,d), \end{array} \qquad \text{(R-OPT}_p\text{)}$$

where $1 \leq p \leq \infty$. We indicated in the introduction of this thesis that it is virtually impossible to actually solve (R-OPT$_p$) numerically. The best thing one can hope for in practice is to find some $n_0 \in L^\infty_{a,b}(0,d)$ such that $\|R_{n_0} - R_{\text{desired}}\|_{L^p(I)}$ is close to $\inf_{n \in L^\infty_{a,b}(0,d)} \|R_n - R_{\text{desired}}\|_{L^p(I)}$. Even proving that the infimum is a minimum, i.e., that (R-OPT$_p$) has a solution, requires several of our new results from this chapter.

Theorem 3.20. *For every $1 \leq p \leq \infty$ the optimization problem* (R-OPT$_p$) *has a solution $n \in L^\infty_{a,b}(0,d)$, i.e., there is $n_0 \in L^\infty_{a,b}(0,d)$ such that $\|R_{n_0} - R_{\text{desired}}\|_{L^p(I)} \leq \|R_n - R_{\text{desired}}\|_{L^p(I)}$ for all $n \in L^\infty_{a,b}(0,d)$.*

Proof. We write the optimization problem in terms of the permittivity ϵ, i.e., we consider

$$\begin{array}{ll} \text{minimize} & J_p(\epsilon) \\ \text{subject to} & \epsilon \in L^\infty_{a^2,b^2}(0,d), \end{array} \qquad (\epsilon\text{-OPT}_p)$$

where

$$J_p(\epsilon) = \|R_{\sqrt{\epsilon}} - R_{\text{desired}}\|_{L^p(I)}.$$

Obviously, problems (ϵ-OPT$_p$) and (R-OPT$_p$) are equivalent, i.e., it suffices to show that (ϵ-OPT$_p$) has at least one solution.

The set $L^\infty_{a,b}(0,d)$ is (sequentially) weak* compact in $L^\infty(0,d)$ (see, e.g., [17, Proposition 2.2]). Thus we have the existence of a minimum if we can show that J_p is (sequentially) weak* continuous on $L^\infty_{a,b}(0,d)$. So let $(\epsilon_j) \subset L^\infty_{a,b}(0,d)$ with $\epsilon_j \overset{*}{\rightharpoonup} \epsilon$.

Case 1: $1 \leq p < \infty$. By Corollary 3.12, $R_{\sqrt{\epsilon_j}}(k) \to R_{\sqrt{\epsilon}}(k)$ pointwise in $k \in \mathbb{R}$. Moreover, $|R_{\sqrt{\epsilon_j}}(k)| \leq 1$ by Theorem 3.5(b). By dominated convergence we then have $\|R_{\sqrt{\epsilon_j}} - R_{\text{desired}}\|_{L^p(I)} \to \|R_{\sqrt{\epsilon}} - R_{\text{desired}}\|_{L^p(I)}$, i.e., J_p is (sequentially) weak* continuous on $L^\infty_{a,b}(0,d)$.

Case 2: $p = \infty$. Let $\eta > 0$ be arbitrary. By the bound on $R'(k)$ from Theorem 3.6 we can find $\delta > 0$ such that for every $\epsilon \in L^\infty_{a,b}(0,d)$

$$\sup_{|k-\widetilde{k}|<\delta} |R_{\sqrt{\epsilon}}(k) - R_{\sqrt{\epsilon}}(\widetilde{k})| \leq \eta/3.$$

Now pick finitely many points $k_l \in I$ such that for all l

$$\min_{\widetilde{l} \neq l} |k_l - k_{\widetilde{l}}| \leq 2\delta.$$

By Corollary 3.12 we can choose $j_0 \in \mathbb{N}$ large enough so that for all $j \geq j_0$ and for all of the finitely many k_l

$$|R_{\sqrt{\epsilon_j}}(k_l) - R_{\sqrt{\epsilon}}(k_l)| \leq \eta/3.$$

For $k \in I$ we can then pick $k_l \in I$ such that $|k - k_l| \le \delta$, and therefore, for $j \ge j_0$,

$$\begin{aligned}
&|R_{\sqrt{\epsilon_j}}(k) - R_{\sqrt{\epsilon}}(k)| \\
&\quad \le |R_{\sqrt{\epsilon_j}}(k) - R_{\sqrt{\epsilon_j}}(k_l)| + |R_{\sqrt{\epsilon_j}}(k_l) - R_{\sqrt{\epsilon}}(k_l)| + |R_{\sqrt{\epsilon}}(k_l) - R_{\sqrt{\epsilon}}(k)| \\
&\quad \le \eta/3 + \eta/3 + \eta/3 = \eta,
\end{aligned}$$

i.e.,

$$\|R_{\sqrt{\epsilon_j}} - R_{\sqrt{\epsilon}}\|_{L^\infty(I)} \le \eta.$$

This proves (sequential) weak* continuity of J_∞ and therefore the theorem.

□

3.5 Further remarks

If the refractive profile n is *smooth*, much more is known. Especially, the *inverse problem* has been studied. For example, for n in certain spaces X of smooth functions the corresponding range of reflection coefficients, $\{R_n : n \in X\}$, has been characterized, and algorithms to reconstruct n from a given reflection coefficient are known. For smooth n one can apply a variable transformation called *Liouville transformation*. It transforms the Helmholtz equation (3.1),

$$u''(x) + k^2 n^2(x) u(x) = 0,$$

either into a variant of the Helmholtz equation [56, 57, 58] or the Schrödinger equation [8]. For example, to obtain the variant of the Helmholtz equation, one sets $t(x) = \int_0^x n(s)\,\mathrm{d}s$, $\gamma(t) = n(x(t))$ and $v(t) = u(x(t))$. Then (3.1) is transformed into

$$v''(t) + \alpha(t) v'(t) + k^2 v(t) = 0,$$

where $\alpha(t) = \frac{\gamma'(t)}{\gamma(t)}$. It has been shown that the scattering operator $s : \alpha \mapsto R$, where R is the reflection coefficient corresponding to α, is bijective as a mapping from $L^2(0,\infty)$ to the Hardy space $\mathcal{H}^E(\mathbb{C}^+)$, where

$$\mathcal{H}^E(\mathbb{C}^+) = \left\{ f : f \text{ analytic on } \mathbb{C}^+,\ \sup_{b>0} E(f(\cdot + ib)) < \infty,\ f(-\bar{k}) = \overline{f(k)} \right\},$$

and

$$E(f) = \int_{-\infty}^{\infty} -\log(1 - |f(x)|^2)\,\mathrm{d}x.$$

Moreover, an algorithm to reconstruct α (and thus n) from $R \in \mathcal{H}^E(\mathbb{C}^+)$ is known, the *layer-stripping method*. Similarly, in the case of the Schrödinger equation, there are results concerning the range of reflection coefficients and methods to reconstruct the refractive profile from a given reflection coefficient in this range [1, 8, 13, 14, 18, 21, 37, 47].

It has been attempted in [11] to use these results for the mirror design problem, but it has turned out that the practical value is limited. First and foremost, the involved spaces do not fit. For example, refractive profiles n corresponding to $R \in \mathcal{H}^E(\mathbb{C}^+)$ (and thus $\alpha \in L^2(0,\infty)$) need *not* satisfy $n|_{\mathbb{R}\setminus[0,d]} = 1$ for some $d > 0$ and may therefore not be physically realizable. Although there is a characterization of $s(L^2(0,B))$ with $B > 0$ [57, Section 3] (i.e., the set of reflection coefficients corresponding to α with support in $(0,B)$), this is not helpful, either. In order to make use of this, we would need to approximate a desired reflection coefficient R_{desired} by some $R \in s(L^2(0,B))$. It is not quite obvious how this could be accomplished. For example, $s(L^2(0,B))$ is not convex.

Moreover, materials that are typically used for optical interference coatings cover only a very limited range of refractive indices. This is why we decided to use the space $L^\infty_{a,b}(0,d)$ for refractive profiles at the beginning of this chapter. If we take $\alpha \in L^2(0,B)$, then the corresponding refractive profile n may still take values that are too large or too small.

Finally, even if we are lucky enough to obtain a physically realizable refractive profile from a reconstruction algorithm, further optimization is necessary. Although the refractive index of typical coating materials is only slightly frequency-dependent, this needs to be included in a final optimization step. The reconstruction algorithms always yield smooth refractive profiles n. For optimization, n has to be approximated by a refractive profile consisting of many very thin layers (i.e., a step function). The number of layers necessary to represent n accurately may be a few thousand. In contrast, the more common binary structures such as in Figure 0.1 usually consist of at *most* 100 or 200 layers, which makes final optimization much easier.

Chapter 4

Constrained Optimization in Hardy Spaces: Theory

> It seems that physicists do not object to rigorous proofs provided they are rather short and simple.
>
> E. C. Titchmarsh, *Eigenfunction Expansions, Part II*

In this chapter we consider the following problem.

Problem 4.1. *Let* $g \in C(\partial\mathbb{D})$ *with* $g > 0$, $K \subset \partial\mathbb{D}$ *closed with positive measure and* $\varphi \in C(K)$ *such that* $|\varphi| \le g$ *on* K. *Moreover, let* $p \in [1, \infty]$. *We are interested in the optimization problem*

$$\begin{aligned} \text{minimize} \quad & \|f - \varphi\|_{L^p(K)} \\ \text{subject to} \quad & f \in E, \\ & |f| \le g \quad \text{on } \partial\mathbb{D}. \end{aligned} \tag{OPT_p}$$

Here, E *is either the space* $H^\infty(\mathbb{D})$ *or the space* $\mathcal{A}(\mathbb{D})$. *In the first case we denote the problem by* (H-OPT_p), *and in the second case we denote the problem by* ($\mathcal{A}\text{-OPT}_p$).

In some cases it may also be desirable to admit more general g. We prove most of the theorems in this and the next chapter for continuous g and remark only afterwards whether the assumptions on g can be weakened, how this possibly changes the theorem, and what has to be changed about the proof.

The motivation for considering the above problem is the following. In the last chapter and in the introduction of this thesis we mentioned that it is extremely hard to find a global optimum of the problem

$$\begin{aligned} \text{minimize} \quad & \|R_n - R_{\text{desired}}\|_{L^p(I)} \\ \text{subject to} \quad & n \in L^\infty_{a,b}(0, d). \end{aligned} \tag{R-OPT_p}$$

A general drawback of all available optimization methods is that they do not give the user information on how close the current solution is to a global optimum: If the current solution *almost* satisfies some desired specifications arising from an application, there is no way to tell whether an algorithm might just need another hour of CPU time to find an acceptable solution, or whether even the global optimum of the problem (R-OPT$_p$) does not satisfy the desired specifications.

Our goal is now to compute a *bound* for the minimum of (R-OPT$_p$). The idea is to replace the search space of realizable reflection coefficients $\{R_n : n \in L^\infty_{a,b}(0,d)\}$ by a larger (but not too large) space which has nicer properties, that is, for which it is easier to find a global (with respect to the bigger space) minimum of the objective function. We saw in the last chapter that the space of realizable reflection coefficients R is rather restricted: It is necessary that $R \in H^\infty(\mathbb{C}^+)$ (causality principle) with $\|R\|_{H^\infty(\mathbb{C}^+)} \leq 1$ (no gain of energy), see Theorem 3.19. So if we replace $\{R_n : n \in L^\infty_{a,b}(0,d)\}$ by the convex set $\{f \in H^\infty(\mathbb{C}^+) : \|f\|_{H^\infty(\mathbb{C}^+)} \leq 1\}$ and use the mapping from Theorem 2.12 to transport everything from $H^\infty(\mathbb{C}^+)$ to $H^\infty(\mathbb{D})$, we end up with the problem (H-OPT$_p$) (with $g \equiv 1$, φ is the desired reflection coefficient transported to $H^\infty(\mathbb{D})$, and K is I transported to the circle).

By allowing general $g \in C(\partial\mathbb{D})$ instead of $g \equiv 1$, we can model additional restrictions on the reflection coefficient arising from applications. For example, in the design of dispersion-compensating mirrors for the compression of laser pulses one might be interested in having a pump window [51], that is, a frequency interval where the reflection coefficient is small. We can incorporate this into our model by choosing g to be small in that particular interval.

Moreover, the reflection coefficient R must satisfy the reality condition $\overline{R(k)} = R(-\overline{k})$, see Theorem 3.5. However, we do not incorporate this into (H-OPT$_p$) since we will see in Section 4.3 that under symmetry assumptions on φ, g and K the solution of (H-OPT$_p$) also satisfies real symmetry.

We begin in Section 4.1 by proving existence ($1 \leq p \leq \infty$) and uniqueness ($1 < p < \infty$) for (H-OPT$_p$). In Section 4.2 we show that the solution of (H-OPT$_p$) satisfies a remarkable extremal property. From this we also deduce uniqueness for the case $p = \infty$ and, under the assumption that $K \neq \partial\mathbb{D}$, the case $p = 1$. As just mentioned, Section 4.3 is devoted to symmetry properties. In Section 4.4 we show that the minimum of (H-OPT$_p$) ($1 \leq p < \infty$) and the infimum of ($\mathcal{A}$-OPT$_\infty$) can be approximated by polynomials.

This is important because in numerical computations we can only work with polynomials. In Section 4.5 we show that if K is nice enough, then the infimum of ($\mathcal{A}$-OPT$_\infty$) is equal to the minimum of (H-OPT$_\infty$). This also makes (H-OPT$_\infty$) accessible for numerical solution.

4.1 Existence ($1 \le p \le \infty$) and uniqueness ($1 < p < \infty$)

The first question we have to address is of course whether (H-OPT$_p$) has a solution and, if it has a solution, whether this solution is unique.

Theorem 4.2. (H-OPT$_p$) *has a solution,* $1 \le p \le \infty$.

Proof. Let $(f_n) \subset H^\infty(\mathbb{D})$ with $|f_n| \le g$ be a minimizing sequence*, i.e., $\|f_n - \varphi\|_{L^p(K)} \to \inf_{f \in H^\infty(\mathbb{D}), |f| \le g \text{ on } \partial\mathbb{D}} \|f - \varphi\|_{L^p(K)}$. It is always possible to choose such a sequence since the set of feasible solutions $\{f \in H^\infty(\mathbb{D}) : |f| \le g\}$ is non-empty: It contains $f \equiv 0$. Obviously, $\|f_n\|_{H^\infty(\mathbb{D})}$ is bounded independent of n (by $\|g\|_{L^\infty(\partial\mathbb{D})}$). It follows that (f_n) is a normal family (see, e.g., [49]), i.e., it contains a subsequence (which we also call (f_n)) that converges uniformly on compact subsets of $\mathbb{D}$ to some function $f \in H^\infty(\mathbb{D})$.

We use the subscript r to denote the Poisson integral of functions on the circle, that is, $f_r(\mathrm{e}^{i\vartheta}) = f(r\mathrm{e}^{i\vartheta}) = \frac{1}{2\pi}\int_{-\pi}^{\pi} f(\mathrm{e}^{it}) P_r(\vartheta - t)\,\mathrm{d}t$, where P_r is the Poisson kernel for the disk. Using Theorem 2.3, we have

$$\begin{aligned}
\|f - \varphi\|_{L^p(K)} &= \|\mathbf{1}_K(f-\varphi)\|_{L^p(\partial\mathbb{D})} \\
&= \lim_{r \nearrow 1} \|(\mathbf{1}_K(f-\varphi))_r\|_{L^p(\partial\mathbb{D})} \\
&= \lim_{r \nearrow 1} \lim_{n \to \infty} \underbrace{\|(\mathbf{1}_K(f_n-\varphi))_r\|_{L^p(\partial\mathbb{D})}}_{\le \|\mathbf{1}_K(f_n-\varphi)\|_{L^p(\partial\mathbb{D})} \text{ by Thm. 2.3(b)}} \\
&\le \liminf_{n\to\infty} \|\mathbf{1}_K(f_n-\varphi)\|_{L^p(\partial\mathbb{D})} \\
&= \liminf_{n\to\infty} \|f_n - \varphi\|_{L^p(K)}.
\end{aligned}$$

For the third equality we used that f_n converges uniformly to f on compact subsets of $\mathbb{D}$ and the fact that f_n is the Poisson integral of its boundary values (Theorem 2.4).

It remains to show that f is feasible, that is, $|f| \le g$. Because f_n is feasible, $g - |f_n| \ge 0$ on $\partial\mathbb{D}$. Since the Poisson kernel is nonnegative, the Poisson

*In this and the next chapter, we use n as an index, and not for refractive profiles. From this chapter on, we do not deal with refractive profiles any more, so no confusion will arise.

integrals $(g - |f_n|)_r$ are nonnegative. Due to uniform convergence $(g - |f|)_r$ is also nonnegative. From the pointwise convergence $(g - |f|)_r \to g - |f|$ a.e. as $r \nearrow 1$ it follows that $g - |f|$ is nonnegative.

Together, it follows that f is a solution of (H-OPT$_p$). □

Uniqueness is especially easy in the case $1 < p < \infty$.

Theorem 4.3. (H-OPT$_p$) *is uniquely solvable for* $1 < p < \infty$.

Proof. The theorem follows directly from the fact that the norm on $L^p(K)$ is strictly convex for $1 < p < \infty$: Assume that f_1^* and f_2^* are both solutions of (H-OPT$_p$), i.e., they minimize $\|f - \varphi\|_{L^p(K)}$ over $\{f \in H^\infty(\mathbb{D}) : |f| \leq g \text{ on } \partial\mathbb{D}\}$. Let $\tau^* = \|f_1^* - \varphi\|_{L^p(K)} = \|f_2^* - \varphi\|_{L^p(K)}$ be the minimum. Due to convexity of the latter set, $(f_1^* + f_2^*)/2$ is also feasible. Suppose that $f_1^* \neq f_2^*$. Then due to strict convexity of the norm, $\|(f_1^* + f_2^*)/2 - \varphi\|_{L^p(K)} < (\|f_1^* - \varphi\|_{L^p(K)} + \|f_2^* - \varphi\|_{L^p(K)})/2 = \tau^*$. This is a contradiction, so the solution must be unique. □

We wish to point out that so far we have not used the assumption that $|\varphi| \leq g$ on K. We will see in the next section that this assumption ensures uniqueness in the cases $p = \infty$ and, if additionally $K \neq \partial\mathbb{D}$, $p = 1$.

4.2 Extremal properties and uniqueness ($1 \leq p \leq \infty$)

The solution of (H-OPT$_p$) satisfies a remarkable extremal property.

Theorem 4.4. *Let f^* be a solution of* (H-OPT$_p$) *and* $\tau^* = \|f^* - \varphi\|_{L^p(K)} > 0$. *If $1 \leq p < \infty$, then for almost all* $\mathrm{e}^{i\vartheta} \in \partial\mathbb{D} \setminus K$

$$|f^*(\mathrm{e}^{i\vartheta})| = g(\mathrm{e}^{i\vartheta}).$$

If $p = \infty$, then, for almost all $\mathrm{e}^{i\vartheta} \in \partial\mathbb{D}$, $f^*(\mathrm{e}^{i\vartheta})$ *is on the boundary of the set*

$$S(\vartheta, \tau^*) = \{z \in \mathbb{C} : |z| \leq g(\mathrm{e}^{i\vartheta}),\ |z - \varphi(\mathrm{e}^{i\vartheta})| \leq \tau^* \text{ if } \mathrm{e}^{i\vartheta} \in K\}.$$

Moreover, for $1 < p \leq \infty$, the solution of (H-OPT$_p$) *is unique. If $K \neq \partial\mathbb{D}$, then the solution of* (H-OPT$_p$) *is also unique for $p = 1$.*

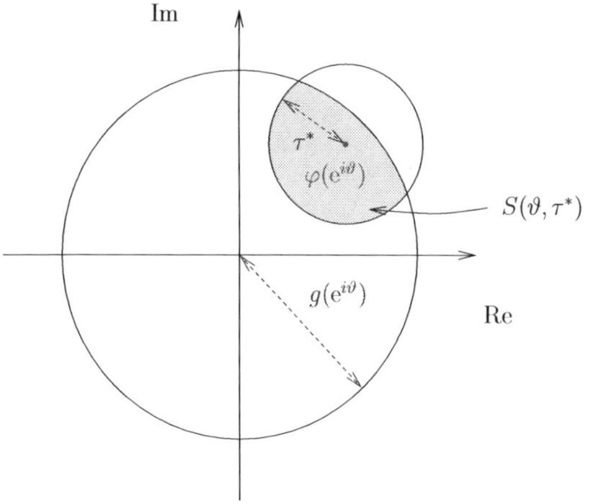

Figure 4.1: Theorem 4.4 says that if f^* is a solution of (H-OPT$_\infty$), then for a.a. $e^{i\vartheta} \in K$, $f^*(e^{i\vartheta})$ is on the boundary of the intersection of the above two disks. Moreover, for a.a. $e^{i\vartheta} \in \partial\mathbb{D} \setminus K$, $f^*(e^{i\vartheta})$ is on the boundary of the disk $\{|z| \le g(e^{i\vartheta})\}$.

Remark 4.5. *If $\tau^* = 0$, then $\varphi = f^*|_K$ for any solution f^* of* (H-OPT$_\infty$). *In this case, uniqueness follows from Theorem 2.6: If $\widetilde{f}$ is another solution, then $\widetilde{f} = f^*$ on K, so $\widetilde{f} = f^*$ on $\partial\mathbb{D}$. However, it is not hard to see that the implication of Theorem 4.4 need* not *hold true in general if $\tau^* = 0$: Given some closed $K \subset \partial\mathbb{D}$ with $\lambda(K) > 0$ and $\lambda(\partial\mathbb{D} \setminus K) > 0$, take any $\varphi \in C(K)$ such that φ is the restriction to K of some nonzero function $\psi \in \mathcal{A}(\mathbb{D})$ and use $g \equiv 2\|\psi\|_{H^\infty(\mathbb{D})}$. The solution of* (H-OPT$_p$) *with these K, φ and g is just $f^* = \psi$. But clearly $f^* < g$, and in the case $p = \infty$, $f^*(e^{i\vartheta})$ is* not *on the boundary of $S(\vartheta, 0)$ for almost all* $e^{i\vartheta} \in \partial\mathbb{D}$.

Remark 4.6. *Theorem 4.4 is a variant of a well-known result for the* Nehari problem *(see, e.g., [25, Chapter IV, Theorem 1.3]): For a given function $\varphi \in L^\infty(\partial\mathbb{D})$ there exists a unique best approximation f^* in $H^\infty(\mathbb{D})$ (without constraints), and $|f^* - \varphi|$ is a.e. constant. This means that for $\tau^* = |f^* - \varphi| = \inf_{f \in H^\infty(\mathbb{D})} \|f - \varphi\|_{L^\infty(\partial\mathbb{D})}$, $f^*(e^{i\vartheta})$ is on the boundary of the set $\{z \in \mathbb{C} : |z - \varphi(e^{i\vartheta})| \le \tau^*\}$ for almost all* $e^{i\vartheta} \in \partial\mathbb{D}$.

Various generalizations of the Nehari problem that have special cases in common with (H-OPT$_p$) *have been considered. For example, if $p = \infty$, g is constant on $\partial\mathbb{D} \setminus K$ and g is so large on K that the constraint $|f| \le g$ is* *not active on K, then* (H-OPT$_p$) *is a special case of a problem that has been studied in the context of system identification, see [5], and also [4, 6]. An extremal property similar to the one from Theorem 4.4 can be shown for the solution of this problem.*

Another generalization of the Nehari problem arising in H^∞ control theory has been studied quite extensively in the literature, see, e.g., [30] or [31] and the references therein: Given a performance function $\Gamma : \partial\mathbb{D} \times \mathbb{C} \to [0, \infty)$ *one is interested in minimizing* $\|\Gamma(\cdot, f(\cdot))\|_{L^\infty(\partial\mathbb{D})}$ *over* $f \in H^\infty(\mathbb{D})$. *When* $\Gamma(\mathrm{e}^{i\vartheta}, z) = |\varphi(\mathrm{e}^{i\vartheta}) - z|$ *for some* $\varphi \in L^\infty(\partial\mathbb{D})$, *then this is the Nehari problem. Under certain assumptions on* Γ *one can prove that there is a unique minimizer* f^*, *and that* $\Gamma(\mathrm{e}^{i\vartheta}, f^*(\mathrm{e}^{i\vartheta}))$ *is a.e. constant. This means that, with* $\tau^* = \inf_{f \in H^\infty(\mathbb{D})} \|\Gamma(\cdot, f(\cdot))\|_{L^\infty(\partial\mathbb{D})}$, $f^*(\mathrm{e}^{i\vartheta})$ *is on the boundary of the set* $\{z \in \mathbb{C} : \Gamma(\mathrm{e}^{i\vartheta}, z) \leq \tau^*\}$ *for almost all* $\mathrm{e}^{i\vartheta} \in \partial\mathbb{D}$. *Depending on the assumptions one puts on* Γ, *there are various ways to prove this (see, e.g., [29, 30, 31, 33]). Indeed, the main idea of our proof of Theorem 4.4 is from the proof of [29, Theorem 1].*

As it is common in convex optimization, we are going to use the Hahn-Banach Theorem to prove Theorem 4.4. Before we can do this, we need an auxiliary result.

Let

$$\mathscr{S}_p = \{f \in L^\infty(\partial\mathbb{D}) : |f| \leq g,\ \|f - \varphi\|_{L^p(K)} \leq \tau^*\},$$

where τ^* is the minimum of (H-OPT$_p$).

Lemma 4.7. *Let $1 \leq p \leq \infty$ and assume that $\tau^* > 0$.*

(a) *The set $\mathscr{S}_p$ is convex.*

(b) *The interior of $\mathscr{S}_p$ is non-empty and disjoint from $\mathcal{A}(\mathbb{D})$.*

(c) *Every element of $\mathscr{S}_p$ is a pointwise limit of functions from $\mathscr{S}_p \cap C(\partial\mathbb{D})$.*

Proof. There is not much to show for (a), because it is immediate from the definition that $\mathscr{S}_p$ is convex. Moreover, it is clear that the interior of $\mathscr{S}_p$ cannot contain any function from $\mathcal{A}(\mathbb{D})$: If there were such a function f, we would have $\|f - \varphi\|_{L^p(K)} < \tau^*$, contradicting the definition of τ^*. This is the second part of (b).

By Tietze's Extension Theorem [49, Theorem 20.4], φ can be extended to a function that is continuous on $\partial\mathbb{D}$. We also denote this extension by φ.

We arrange it so that $|\varphi| \le g$ on $\partial\mathbb{D}$. Now let $\epsilon = \tau^*/(2\|\mathbf{1}\|_{L^p(K)})$ and define

$$a(\mathrm{e}^{i\vartheta}) = \begin{cases} \varphi(\mathrm{e}^{i\vartheta}), & |\varphi(\mathrm{e}^{i\vartheta})| \le g(\mathrm{e}^{i\vartheta}) - \epsilon, \\ \frac{\varphi(\mathrm{e}^{i\vartheta})}{|\varphi(\mathrm{e}^{i\vartheta})|}(g(\mathrm{e}^{i\vartheta}) - \epsilon), & \text{otherwise.} \end{cases}$$

It is straightforward to prove that $\|a - \varphi\|_{L^\infty(\partial\mathbb{D})} \le \epsilon$ and $|a| \le g - \epsilon$. Let $v \in L^\infty(\partial\mathbb{D})$ with $\|v\|_{L^\infty(\partial\mathbb{D})} \le \epsilon$. Then $\|(a+v) - \varphi\|_{L^p(K)} \le \|2\epsilon\mathbf{1}\|_{L^p(K)} = \tau^*$ and $|a+v| \le g$, i.e., $a+v \in \mathscr{S}_p$. Because τ^* is positive by assumption, a lies in the interior of $\mathscr{S}_p$. This finishes the proof of (b).

The proof of (c) is not particularly hard, but a little more technical. We first consider the case $1 \le p < \infty$. Let $f \in \mathscr{S}_p$. Then there is a sequence $(\widetilde{f}_n) \subset C(\partial\mathbb{D})$ with $\|\widetilde{f}_n\|_{L^\infty(\partial\mathbb{D})} \le \|f\|_{L^\infty(\partial\mathbb{D})}$ such that $\widetilde{f}_n \to f$ a.e. (see, e.g., [49, Chapter 2]). By dominated convergence, $\|\widetilde{f}_n - \varphi\|_{L^p(K)} \to \|f - \varphi\|_{L^p(K)} \le \tau^*$. Now set

$$\widetilde{f}_n^1 = \varphi + (\widetilde{f}_n - \varphi)\frac{\|f - \varphi\|_{L^p(K)}}{\|\widetilde{f}_n - \varphi\|_{L^p(K)}}.$$

Then $\widetilde{f}_n^1 \in C(\partial\mathbb{D})$, $\widetilde{f}_n^1 \to f$ a.e., and moreover $\|\widetilde{f}_n^1 - \varphi\|_{L^p(K)} = \|f - \varphi\|_{L^p(K)} \le \tau^*$. However, it may not hold true that $|\widetilde{f}_n^1| \le g$. We therefore define functions f_n by

$$f_n(\mathrm{e}^{i\vartheta}) = \begin{cases} \widetilde{f}_n^1(\mathrm{e}^{i\vartheta}), & |\widetilde{f}_n^1(\mathrm{e}^{i\vartheta})| \le g(\mathrm{e}^{i\vartheta}), \\ \varphi(\mathrm{e}^{i\vartheta}) + \mu_n(\mathrm{e}^{i\vartheta})(\widetilde{f}_n^1(\mathrm{e}^{i\vartheta}) - \varphi(\mathrm{e}^{i\vartheta})), & \text{otherwise,} \end{cases}$$

where μ_n is a function on $\partial\mathbb{D}$ such that, if we are in the second case of the above definition, then $|f_n(\mathrm{e}^{i\vartheta})| = g(\mathrm{e}^{i\vartheta})$ as in Figure 4.2. Concretely, we set $\mu_n(\mathrm{e}^{i\vartheta}) = 1/p_\vartheta(\widetilde{f}_n^1(\mathrm{e}^{i\vartheta}))$, where p_ϑ is the Minkowski functional

$$p_\vartheta(z) = \inf\{t > 0 : |\varphi(\mathrm{e}^{i\vartheta}) + t^{-1}(z - \varphi(\mathrm{e}^{i\vartheta}))| \le g(\mathrm{e}^{i\vartheta})\}.$$

Then f_n is continuous, $|f_n| \le g$, and $f_n \to f$ pointwise a.e. Moreover, if $|\widetilde{f}_n^1(\mathrm{e}^{i\vartheta})| > g(\mathrm{e}^{i\vartheta})$, then $p_\vartheta(\widetilde{f}_n^1(\mathrm{e}^{i\vartheta})) > 1$, and therefore

$$|f_n(\mathrm{e}^{i\vartheta}) - \varphi(\mathrm{e}^{i\vartheta})| = \left|\frac{\widetilde{f}_n^1(\mathrm{e}^{i\vartheta}) - \varphi(\mathrm{e}^{i\vartheta})}{p_\vartheta(\widetilde{f}_n^1(\mathrm{e}^{i\vartheta}))}\right| \le |\widetilde{f}_n^1(\mathrm{e}^{i\vartheta}) - \varphi(\mathrm{e}^{i\vartheta})|.$$

If follows that $\|f_n - \varphi\|_{L^p(K)} \le \|\widetilde{f}_n^1 - \varphi\|_{L^p(K)} \le \tau^*$, so $f_n \in \mathscr{S}_p \cap C(\partial\mathbb{D})$. This proves (c) for $1 \le p < \infty$.

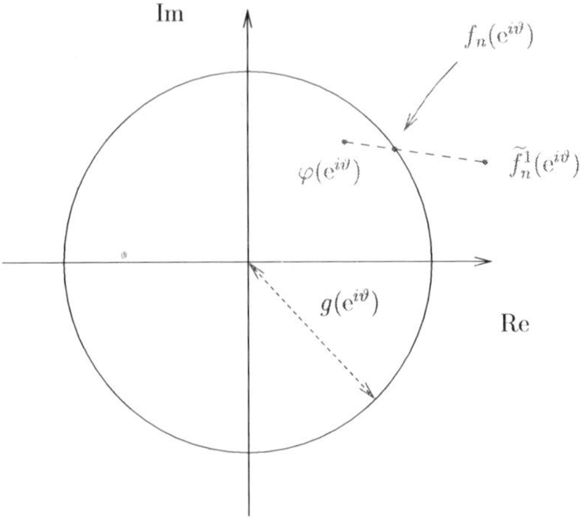

Figure 4.2: Definition of the functions f_n.

It remains to consider the case $p = \infty$. Let $f \in \mathscr{S}_\infty$. Then we have $\|f - \varphi\|_{L^\infty(K)} \leq \tau^*$. As before, there is a sequence $(f_n^K) \subset C(K)$ such that $f_n^K \to f$ pointwise a.e. on K and $\|f_n^K - \varphi\|_{L^\infty(K)} \leq \tau^*$, and we can arrange it so that $|f_n^K| \leq g$ on K. By Tietze's Extension Theorem, every f_n^K can be extended to a function that is continuous on $\partial\mathbb{D}$. We also denote this extension by f_n^K, and we arrange it so that $|f_n^K| \leq g$ on $\partial\mathbb{D}$. Similarly, there is a sequence $(f_n^{\partial\mathbb{D}}) \subset C(\partial\mathbb{D})$ such that $f_n^{\partial\mathbb{D}} \to f$ pointwise a.e. on $\partial\mathbb{D}$, and we arrange it so that $|f_n^{\partial\mathbb{D}}| \leq g$ on $\partial\mathbb{D}$.

Now let $U_n \subset \partial\mathbb{D}$ be open with $K \subset U_n$ and $\text{meas}(U_n \setminus K) \leq \frac{1}{2^n}$, where meas denotes Lebesgue measure on the circle. Regularity properties of Lebesgue measure ensure that this is always possible [49]. By Urysohn's Lemma there is $h_n \in C(\partial\mathbb{D})$ with $h_n \equiv 1$ on K, $h_n \geq 0$ and $\operatorname{supp} h_n \subset U_n$. Set

$$f_n = h_n f_n^K + (1 - h_n) f_n^{\partial\mathbb{D}}.$$

Then $f_n \in C(\partial\mathbb{D})$, $f_n \to f$ a.e., and $|f_n| \leq g$ on $\partial\mathbb{D}$. Moreover, $f_n = f_n^K$ on K, whence $\|f_n - \varphi\|_{L^\infty(K)} = \|f_n^K - \varphi\|_{L^\infty(K)} \leq \tau^*$. Therefore, $f_n \in \mathscr{S}_\infty$. This proves (c) for $p = \infty$. □

We are now ready to prove Theorem 4.4. Our proof goes along the lines of the proof of [29, Theorem 2].

Proof of Theorem 4.4. By the Hahn-Banach Theorem and properties (a) and (b) of Lemma 4.7, there is a nonzero $\lambda \in C(\partial\mathbb{D})^*$ such that

$$\operatorname{Re}\lambda(\mathscr{S}_p \cap C(\partial\mathbb{D})) \leq \operatorname{Re}\lambda(\mathcal{A}(\mathbb{D})). \tag{4.1}$$

Because $\mathcal{A}(\mathbb{D})$ is a linear space, we must have either $\operatorname{Re}\lambda(\mathcal{A}(\mathbb{D})) = \mathbb{R}$ or $\operatorname{Re}\lambda(\mathcal{A}(\mathbb{D})) = 0$. But because of (4.1), $\operatorname{Re}\lambda(\mathcal{A}(\mathbb{D}))$ is bounded from below, whence

$$\operatorname{Re}\lambda(\mathcal{A}(\mathbb{D})) = 0. \tag{4.2}$$

Therefore, there is a nonzero $l \in H_0^1(\mathbb{D}) = \{f \in H^1(\mathbb{D}) : f(0) = 0\}$ such that

$$\lambda(f) = \int_{-\pi}^{\pi} f(\mathrm{e}^{i\vartheta})l(\mathrm{e}^{i\vartheta})\,\mathrm{d}\vartheta$$

for all $f \in C(\partial\mathbb{D})$, see, e.g., [25, Chapter IV]. Using the right hand side of the above equation, we can extend λ to all of $L^\infty(\partial\mathbb{D})$.

If $f \in \mathscr{S}_p$, then by property (c) of Lemma 4.7, there is a sequence $(f_n) \subset \mathscr{S}_p \cap C(\partial\mathbb{D})$ such that $f_n \to f$ pointwise a.e. Because (f_n) is bounded by g, dominated convergence yields $\lambda(f_n) \to \lambda(f)$. It follows from (4.1) and (4.2) that

$$\operatorname{Re}\lambda(f) \leq 0 \text{ for all } f \in \mathscr{S}_p. \tag{4.3}$$

Further, if $f \in H^\infty(\mathbb{D})$, then f is the pointwise limit of functions in $\mathcal{A}(\mathbb{D})$. This follows for example from Theorem 2.4. Dominated convergence and (4.2) imply

$$\operatorname{Re}\lambda(f) = 0 \text{ for all } f \in H^\infty(\mathbb{D}). \tag{4.4}$$

We now prove the assertion of the theorem for the case $1 \leq p < \infty$. Let f^* be a solution of (H-OPT$_p$). Then $f^* \in H^\infty(\mathbb{D}) \cap \mathscr{S}_p$. If $\partial\mathbb{D} \setminus K$ has zero measure, there is nothing to show, so we can assume that $\partial\mathbb{D} \setminus K$ has positive measure. Assume to the contrary that it is not true that $|f^*| = g$ a.e. on $\partial\mathbb{D} \setminus K$. Then there are a set $I \subset \partial\mathbb{D} \setminus K$ of positive measure and $\epsilon > 0$ such that $|f^*| + \epsilon \leq g$ on I. Let $h \in L^\infty(\partial\mathbb{D})$ be any function with $\|h\|_{L^\infty(\partial\mathbb{D})} \leq \epsilon$ and $\operatorname{supp} h \subset I$. Then $f^* + h \in \mathscr{S}_p$, and

$$0 \overset{(4.3)}{\geq} \operatorname{Re}\int_{-\pi}^{\pi} \left(f^*(\mathrm{e}^{i\vartheta}) + h(\mathrm{e}^{i\vartheta})\right) l(\mathrm{e}^{i\vartheta})\,\mathrm{d}\vartheta \overset{(4.4)}{=} \operatorname{Re}\int_{-\pi}^{\pi} h(\mathrm{e}^{i\vartheta})l(\mathrm{e}^{i\vartheta})\,\mathrm{d}\vartheta.$$

The same inequality follows for $-h$, ih and $-ih$, whence $\int_{-\pi}^{\pi} h(\mathrm{e}^{i\vartheta})l(\mathrm{e}^{i\vartheta})\,\mathrm{d}\vartheta = 0$ for all $h \in L^\infty(\partial\mathbb{D})$. But then $l = 0$ on I, and Theorem 2.6 implies $l = 0$

on $\partial\mathbb{D}$. This is a contradiction to $l \neq 0$. Therefore, it must hold true that $|f^*| = g$ a.e. on $\partial\mathbb{D} \setminus K$.

The statement for the case $p = \infty$ follows with a similar argument.

From Theorem 4.3 we already know that the solution of (H-OPT$_p$) is unique for $1 < p < \infty$. For $p = \infty$, uniqueness follows from the fact that the sets $S(\vartheta, \tau^*)$ are strictly convex for all ϑ: If f_1^* and f_2^* are both solutions of (H-OPT$_\infty$), then $(f_1^* + f_2^*)/2$ is also a solution of (H-OPT$_\infty$), because the norm $\|\cdot\|_{L^\infty(K)}$ is convex. Because the sets $S(\vartheta, \tau^*)$ are strictly convex and $f_j^*(\mathrm{e}^{i\vartheta})$ is on the boundary of $S(\vartheta, \tau^*)$ for almost all $\mathrm{e}^{i\vartheta}$, $j = 1, 2$, it follows that $f_1^* = f_2^*$. If $p = 1$ and $K \neq \partial\mathbb{D}$, then $\partial\mathbb{D} \setminus K$ is nonempty and open and therefore has positive measure. Uniqueness then follows in the same way from the fact that the sets $\{z \in \mathbb{C} : |z| \leq g(\mathrm{e}^{i\vartheta})\}$ are strictly convex for all $\mathrm{e}^{i\vartheta} \in \partial\mathbb{D} \setminus K$. □

Remark 4.8. *Theorem 4.4 still holds true if we admit more general g in Problem 4.1, for example, if g is continuous up to finitely many jump discontinuities. We only used the continuity of g in the proof of Lemma 4.7(c). In order to prove Theorem 4.4 for this case, one has to adapt that proof. We leave out the details, because they are technical and do not add any insight.*

The following example demonstrates that Theorem 4.4 need *not* hold true if we drop the assumption that $|\varphi| \leq g$ on K.

Example 4.9. *Let $K = \partial\mathbb{D}$, $\varphi(\mathrm{e}^{i\vartheta}) = 2$ and $g(\mathrm{e}^{i\vartheta}) = |2 + \mathrm{e}^{i\vartheta}|$. Then the solution of* (H-OPT$_\infty$) *is not unique, and also the extremal property from Theorem 4.4 is not satisfied. Indeed, because $g(\mathrm{e}^{i\pi}) = 1$, we have $\min_{f \in H^\infty(\mathbb{D}),\, |f| \leq g} \|f - \varphi\|_{L^\infty(K)} \geq 1$. On the other hand, let $f_0(\mathrm{e}^{i\vartheta}) = 1$, $f_1(\mathrm{e}^{i\vartheta}) = 2 + \mathrm{e}^{i\vartheta}$ and $f_\lambda(\mathrm{e}^{i\vartheta}) = \lambda f_0(\mathrm{e}^{i\vartheta}) + (1 - \lambda) f_1(\mathrm{e}^{i\vartheta})$. Then f_λ is feasible for* (H-OPT$_\infty$), *$0 \leq \lambda \leq 1$, and $\|f_\lambda - \varphi\|_{L^\infty(K)} = 1$. Thus, the solution of* (H-OPT$_\infty$) *is not unique. Further, f_λ does not satisfy the extremal property from Theorem 4.4 for $0 < \lambda < 1$.*

The reason why the proof of Theorem 4.4 fails if we drop the assumption $|\varphi| \leq g$ on K is that the set $\mathscr{S}_\infty$ may have empty interior, i.e., $\mathscr{S}_\infty$ may not satisfy property (b) from Lemma 4.7. Indeed, in Example 4.9 this is the case because of the singularity at $\vartheta = \pi$, see Figure 4.3. However, one can show that $\mathscr{S}_\infty$ satisfies the conditions from Lemma 4.7 under additional assumptions, for example, if τ^* satisfies $\tau^* > \sup_{\mathrm{e}^{i\vartheta} \in \partial\mathbb{D}} |\varphi(\mathrm{e}^{i\vartheta})| - g(\mathrm{e}^{i\vartheta})$.

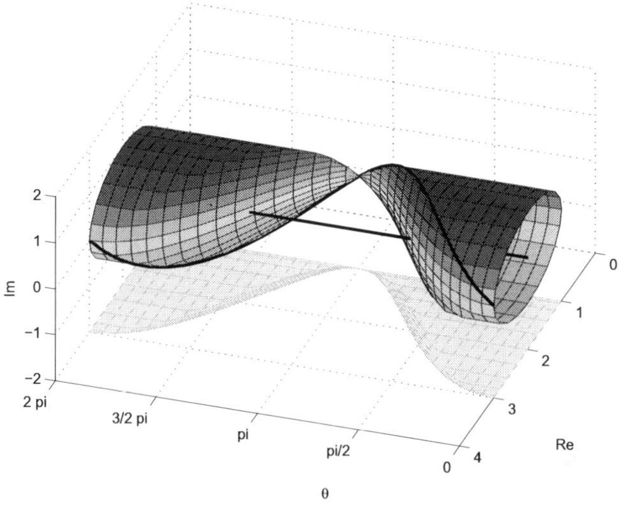

Figure 4.3: In Example 4.9, functions from the set $\mathscr{S}_\infty$ take values in the above tube. The set $\mathscr{S}_\infty$ has empty interior because of the singularity at $\vartheta = \pi$. The straight black line is φ, and the other black curve is f_1.

Remark 4.10. *Under some additional assumptions on g and φ we can obtain a continuity result. Recall that a function f defined on $\partial\mathbb{D}$ is called* Dini continuous *if for some $\epsilon > 0$ it holds that*

$$\int_0^\epsilon \frac{\omega_f(t)}{t}\,\mathrm{d}t < \infty,$$

where

$$\omega_f(\delta) = \sup\{|f(\mathrm{e}^{i\vartheta}) - \mathrm{e}^{it}| : |\vartheta - t| < \delta\}$$

is the modulus of continuity *of f. Moreover, recall that for $f \in L^\infty(\partial\mathbb{D})$ the* essential range of f near $\mathrm{e}^{i\vartheta}$ *is the set*

$$\operatorname{ess\,ran}(f, \mathrm{e}^{i\vartheta}) = \left\{ z \in \mathbb{C} : \begin{array}{l} f^{-1}(B_{\epsilon_1}(z)) \cap \mathrm{e}^{i[\vartheta-\epsilon_2, \vartheta+\epsilon_2]} \textit{ has positive} \\ \textit{Lebesgue measure for all } \epsilon_1, \epsilon_2 > 0 \end{array} \right\}.$$

Here, $B_\epsilon(z) = \{w \in \mathbb{C} : |z - w| < \epsilon\}$ denotes a ball in $\mathbb{C}$.

Assume that g and φ are Dini continuous. Let f^* be the solution of

(H-OPT$_\infty$) *and assume that* $\tau^* = \|f^* - \varphi\|_{L^\infty(K)} > 0$. *Let*

$$\Gamma_1 = \{e^{i\vartheta} \in \partial\mathbb{D} : \operatorname{ess\,ran}(f^*, e^{i\vartheta}) \subset \partial B_{\tau^*}(\varphi(e^{i\vartheta}))\},$$
$$\Gamma_2 = \{e^{i\vartheta} \in \partial\mathbb{D} : \operatorname{ess\,ran}(f^*, e^{i\vartheta}) \subset B_{g(e^{i\vartheta})}(0)\}.$$

By Theorem 4.4 we especially have $\partial\mathbb{D} \setminus K \subset \Gamma_2$. *Using the techniques from Hui [33], one can show that* f^* *is continuous on* Γ_1° *and* Γ_2°, *where the little circle denotes the interior of a set. A result of Chirka [16, Theorem 33] then implies that if* $\varphi \in C^k(K)$ *and* $g \in C^k(\partial\mathbb{D})$, $k \geq 2$, *then* $f^* \in C^{k-1,1-\epsilon}(\Gamma_1^\circ \cup \Gamma_2^\circ \cup \mathbb{D})$ *for any* $\epsilon > 0$.

We do not know whether under the above assumptions f^* *is also continuous on all of* K°. *The difficulty that arises when one tries to apply the techniques from [33] is that for some* $e^{i\vartheta} \in K$ *the boundary of the set* $S(\vartheta, \tau^*)$ *is not an analytic curve.*

4.3 Symmetry

Let us address the question when the solution of (H-OPT$_p$) is real symmetric, that is, $f^*(e^{i\vartheta}) = \overline{f^*(e^{-i\vartheta})}$ for all $e^{i\vartheta} \in \partial\mathbb{D}$. This is important since we know that the reflection coefficient for the Helmholtz equation is real symmetric.

Theorem 4.11. *Assume that* $K = \overline{K}$ *and that* φ *and* g *are real symmetric. If* $1 < p \leq \infty$, *then the solution of* (H-OPT$_p$) *is real symmetric. If* $p = 1$, *then there is a real symmetric solution.*

Proof. Let f^* be a solution of (H-OPT$_p$). From the assumptions it follows that the function $\tilde{f}(e^{i\vartheta}) = \overline{f^*(e^{-i\vartheta})}$ is also a solution of (H-OPT$_p$). If $1 < p \leq \infty$, then uniqueness implies $\tilde{f} = f^*$, i.e., f^* is real symmetric. If $p = 1$, then note that $(f^* + \tilde{f})/2$ is real symmetric, feasible for (H-OPT$_p$) and

$$\|(f^* + \tilde{f})/2 - \varphi\|_{L^\infty(K)} \leq \frac{1}{2}\|f^* - \varphi\|_{L^\infty(K)} + \frac{1}{2}\|\tilde{f} - \varphi\|_{L^\infty(K)} = \|f^* - \varphi\|_{L^\infty(K)}.$$

Thus, $(f^* + \tilde{f})/2$ is a real symmetric solution of (H-OPT$_p$). □

4.4 Approximation by smooth functions, $1 \leq p < \infty$

In numerical computations we cannot work with general functions from $H^\infty(\mathbb{D})$, but only with polynomials. In this section we show that the minimum of (H-OPT$_p$) ($1 \leq p < \infty$) and the infimum of ($\mathcal{A}$-OPT$_\infty$) can be approximated by polynomials (Theorem 4.13). It follows especially that the infimum of ($\mathcal{A}$-OPT$_p$) is equal to the minimum of (H-OPT$_p$) for $1 \le p < \infty$. In Section 4.5 we will show that under additional assumptions this is also true in the case $p = \infty$.

We begin with a lemma which states that functions that are feasible for (H-OPT$_p$) (or ($\mathcal{A}$-OPT$_\infty$)) can be approximated by polynomials that are feasible for (H-OPT$_p$) (or ($\mathcal{A}$-OPT$_\infty$)).

Lemma 4.12. *Let either $1 \le p < \infty$ and $f^* \in H^p(\mathbb{D})$ or $p = \infty$ and $f^* \in \mathcal{A}(\mathbb{D})$. Assume that $|f^*| \le g$ on $\partial\mathbb{D}$. Then there is a sequence $(f_n) \subset \mathcal{A}(\mathbb{D})$ with $|f_n| \le g$ on $\partial\mathbb{D}$ such that*

$$\|f_n - f^*\|_{L^p(\partial\mathbb{D})} \to 0 \qquad \text{as } n \to \infty. \tag{4.5}$$

Furthermore, we may even arrange it for the f_n to be polynomials, that is, to be of the form

$$f_n(\mathrm{e}^{i\vartheta}) = \sum_{k=0}^{N_n-1} \alpha_{N_n,k}\mathrm{e}^{ik\vartheta}. \tag{4.6}$$

If f^* is real symmetric, then we can arrange it for the f_n to be real symmetric, that is, to have real coefficients $\alpha_{N_n,k}$.

Proof. We first show that there is a sequence (f_n) that satisfies (4.5). In the case $p = \infty$ we can simply take $f_n = f^*$, so we only have to consider the case $p < \infty$. For $0 < r < 1$ let f_r^* be the Poisson integral $f_r^*(\mathrm{e}^{i\vartheta}) = \frac{1}{2\pi}\int_{-\pi}^{\pi} f^*(\mathrm{e}^{it})P_r(\vartheta - t)\,\mathrm{d}t$. We then have $f_r^* \in \mathcal{A}(\mathbb{D})$ and $\|f_r^* - f^*\|_{L^p(\partial\mathbb{D})} \to 0$ as $r \nearrow 1$ by Theorem 2.3, but it might not be true that $|f_r^*| \le g$ on $\partial\mathbb{D}$. We are going to construct sequences (r_n) with $r_n \nearrow 1$ and (η_n) with $\eta_n \to 1$ such that $f_n = \eta_n f_{r_n}^*$ satisfies $|f_n| \le g$ on $\partial\mathbb{D}$. It then follows that

$$\begin{aligned}\|f_n - f^*\|_{L^p(\partial\mathbb{D})} &= \|\eta_n f_{r_n}^* - f^*\|_{L^p(\partial\mathbb{D})}\\ &\le \|f_{r_n}^* - f^*\|_{L^p(\partial\mathbb{D})} + |1 - \eta_n|\|f_{r_n}^*\|_{L^p(\partial\mathbb{D})} \to 0\end{aligned}$$

as $n \to \infty$, since $\|f_{r_n}^*\|_{L^p(\partial\mathbb{D})}$ is bounded by $\|f^*\|_{L^p(\partial\mathbb{D})}$ (see Theorem 2.3).

Fix $\epsilon > 0$. Because g is uniformly continuous on $\partial\mathbb{D}$, there is $\delta > 0$ such that $|\vartheta - t| \le \delta$ implies $|g(\mathrm{e}^{i\vartheta}) - g(\mathrm{e}^{it})| \le \epsilon/2$. Furthermore, as $r \nearrow 1$, P_r becomes increasingly concentrated at 0 so that there is $\varrho < 1$ such that for all $r \in [\varrho, 1)$

$$\max_{t\in[-\pi,\pi]\setminus[-\delta,\delta]} |P_r(t)| < (\epsilon/2)\left(\frac{1}{2\pi}\int_{-\pi}^{\pi} |g(\mathrm{e}^{i\vartheta})|\,\mathrm{d}\vartheta\right)^{-1}.$$

Now for $r \in [\varrho, 1)$

$$\begin{aligned}|f_r^*(\mathrm{e}^{i\vartheta})| &= \left|\frac{1}{2\pi}\int_{-\pi}^{\pi} f^*(\mathrm{e}^{i(\vartheta-t)})P_r(t)\,\mathrm{d}t\right| \\ &\le \left|\frac{1}{2\pi}\int_{[-\delta,\delta]} f^*(\mathrm{e}^{i(\vartheta-t)})P_r(t)\,\mathrm{d}t\right| + \left|\frac{1}{2\pi}\int_{[-\pi,\pi]\setminus[-\delta,\delta]} f^*(\mathrm{e}^{i(\vartheta-t)})P_r(t)\,\mathrm{d}t\right| \\ &\le \max_{t\in[\vartheta-\delta,\vartheta+\delta]} |f^*(\mathrm{e}^{it})| + \left(\max_{t\in[-\pi,\pi]\setminus[-\delta,\delta]} |P_r(t)|\right)\left(\frac{1}{2\pi}\int_{-\pi}^{\pi} |f^*(\mathrm{e}^{it})|\,\mathrm{d}t\right).\end{aligned}$$

For the last inequality we used that P_r is positive and $\frac{1}{2\pi}\int_{-\pi}^{\pi} P_r(t)\,\mathrm{d}t = 1$. Since $|f^*| \le g$,

$$|f_r^*(\mathrm{e}^{i\vartheta})| \le \max_{t\in[\vartheta-\delta,\vartheta+\delta]} |g(\mathrm{e}^{it})| + \left(\max_{t\in[-\pi,\pi]\setminus[-\delta,\delta]} |P_r(t)|\right)\left(\frac{1}{2\pi}\int_{-\pi}^{\pi} |g(\mathrm{e}^{it})|\,\mathrm{d}t\right),$$

and therefore

$$|f_r^*(\mathrm{e}^{i\vartheta})| \le (g(\mathrm{e}^{i\vartheta}) + \epsilon/2) + \epsilon/2 = g(\mathrm{e}^{i\vartheta}) + \epsilon.$$

We have just shown that for an arbitrary $\epsilon > 0$ there is $\varrho = \varrho(\epsilon) < 1$ such that for all $r \in [\varrho, 1)$, $|f_r^*| \le g + \epsilon$.

To finish the proof of (4.5), let (ϵ_n) be a sequence of positive real numbers with $\epsilon_n \to 0$. Choose $r_n \ge \varrho(\epsilon_n)$ such that $r_n \nearrow 1$. Then

$$|f_{r_n}^*| \le g + \epsilon_n \le \left(1 + \frac{\epsilon_n}{\min_{\mathrm{e}^{i\vartheta}\in\partial\mathbb{D}} g(\mathrm{e}^{i\vartheta})}\right) g.$$

Set $\eta_n = \left(1 + \frac{\epsilon_n}{\min_{\mathrm{e}^{i\vartheta}\in\partial\mathbb{D}} g(\mathrm{e}^{i\vartheta})}\right)^{-1}$. Then $\eta_n \to 1$, and $|f_n| = |\eta_n f_{r_n}^*| \le g$, and of course f_n is in $\mathcal{A}(\mathbb{D})$. We have already seen that $\|f_n - f^*\|_{L^p(\partial\mathbb{D})} \to 0$ as $n \to \infty$.

It remains to show that there is a sequence of functions that also satisfies (4.6). Since f_n is continuous, there is a polynomial $\widetilde{f}_n$ such that $\|\widetilde{f}_n - f_n\|_{L^\infty(\partial\mathbb{D})} \le \epsilon_n$. Notice that if f^* is real symmetric, then f_n is real symmetric: In the case $p < \infty$ this is due to the symmetry of the Poisson kernel. In the case $p = \infty$ we chose $f_n = f^*$, so f_n is trivially real symmetric. If f_n is real symmetric, we can also choose the polynomial $\widetilde{f}_n$ to be real symmetric. Using the same argument as before, we can find a sequence (η_n) of real numbers with $\eta_n \to 1$ such that $\eta_n\widetilde{f}_n$ satisfies $|\eta_n\widetilde{f}_n| \le g$ on $\partial\mathbb{D}$. Then

$$\|\eta_n\widetilde{f}_n - f_n\|_{L^p(\partial\mathbb{D})} \le \eta_n\|\widetilde{f}_n - f_n\|_{L^p(\partial\mathbb{D})} + (1-\eta_n)\|f_n\|_{L^p(\partial\mathbb{D})} \to 0,$$

and therefore $(\eta_n\widetilde{f}_n)$ satisfies (4.5) and (4.6). □

From the preceding lemma it follows that the minimum of (H-OPT$_p$), $1 \leq p < \infty$, and the infimum of ($\mathcal{A}$-OPT$_\infty$) can be approximated by polynomials.

Theorem 4.13. *If $1 \leq p < \infty$, then there is a sequence (f_n) of polynomials of the form (4.6) with $|f_n| \leq g$ such that*

$$\|f_n - \varphi\|_{L^p(K)} \to \min_{f \in H^\infty(\mathbb{D}),\, |f| \leq g} \|f - \varphi\|_{L^p(K)} \qquad \text{as } n \to \infty.$$

If $p = \infty$, then there is a sequence (f_n) of polynomials of the form (4.6) with $|f_n| \leq g$ such that

$$\|f_n - \varphi\|_{L^\infty(K)} \to \inf_{f \in \mathcal{A}(\mathbb{D}),\, |f| \leq g} \|f - \varphi\|_{L^\infty(K)} \qquad \text{as } n \to \infty.$$

If $K = \overline{K}$ and φ and g are real symmetric, then we can arrange it for the f_n to be real symmetric.

Proof. Use Lemma 4.12. If $1 \leq p < \infty$, then the assertion about real symmetry follows with Theorem 4.11. If $p = \infty$, then use that under the given symmetry assumptions the infimum can always be approximated by real symmetric functions using the argument from the proof of Theorem 4.11. □

If g is not continuous, then the following version of Theorem 4.13 still holds true.

Theorem 4.14. *Let $1 \leq p \leq \infty$. Suppose that instead of $g \in C(\partial\mathbb{D})$ we have $g \in L^\infty(\partial\mathbb{D})$ with $\inf_{e^{i\vartheta} \in \partial\mathbb{D}} g(e^{i\vartheta}) > 0$. Then there is a sequence (f_n) of polynomials of the form (4.6) with $|f_n| \leq g$ such that*

$$\|f_n - \varphi\|_{L^p(K)} \to \inf_{f \in \mathcal{A}(\mathbb{D}),\, |f| \leq g} \|f - \varphi\|_{L^p(K)} \qquad \text{as } n \to \infty.$$

If $K = \overline{K}$ and φ and g are real symmetric, then we can arrange it for the f_n to be real symmetric.

Proof. Fix $\epsilon > 0$. Pick $f^* \in \mathcal{A}(\mathbb{D})$ with $|f^*| \leq g$ and

$$\|f^* - \varphi\|_{L^p(K)} \leq \inf_{f \in \mathcal{A}(\mathbb{D}),\, |f| \leq g} \|f - \varphi\|_{L^p(K)} + \epsilon/2.$$

Let $\widetilde{g} \in C(\partial\mathbb{D})$ such that $|f^*| \leq \widetilde{g} \leq g$ and $\inf \widetilde{g} > 0$. We can for example take $\widetilde{g}(e^{i\vartheta}) = \max\{|f^*(e^{i\vartheta})|, \inf g\}$. By Lemma 4.12 there is a polynomial $\widetilde{f}$ of the form (4.6) such that $|\widetilde{f}| \leq \widetilde{g}$ and $\|\widetilde{f} - f^*\|_{L^p(\partial\mathbb{D})} < \epsilon/2$. Together,

$$\|\widetilde{f} - \varphi\|_{L^p(K)} \leq \|f^* - \varphi\|_{L^p(K)} + \epsilon/2 \leq \inf_{f \in \mathcal{A}(\mathbb{D}),\, |f| \leq g} \|f - \varphi\|_{L^p(K)} + \epsilon.$$

This proves the first part of the theorem. The part about real symmetry can be seen as before. □

4.5 Approximation by smooth functions, $p = \infty$

In the last section we saw that in the case $1 \leq p < \infty$, the infimum of ($\mathcal{A}$-OPT$_p$) is equal to the minimum of (H-OPT$_p$). In this section we show that under additional assumptions this is still true for $p = \infty$. This whole section is devoted to the proof of

Theorem 4.15. *Assume that K is the disjoint union of finitely many intervals of positive length, i.e., $K = \bigcup_{j=1}^{n} K_j$, where $K_j = \mathrm{e}^{i[\lambda_j, \rho_j]}$ for some $\lambda_j < \rho_j$. Then the infimum of ($\mathcal{A}$-OPT$_\infty$) is equal to the minimum of (H-OPT$_\infty$).*

As far as practice is concerned, the assumption that K is the union of finitely many intervals is not strong. It is satisfied in all of the examples coming from practice that we consider in Chapter 5. Moreover, if this additional assertion of Theorem 4.15 is satisfied, then it follows together with Theorem 4.13 that the minimum of (H-OPT$_\infty$) can even be approximated by polynomials.

The proof of Theorem 4.15 is rather technical and lengthy. We divide it into several lemmas. Before we start with the proof, we try to give an idea of the structure.

- Lemmas 4.16 and 4.17 deal with the construction and properties of certain analytic functions ψ_δ mapping $\overline{\mathbb{D}}$ into $\mathbb{D}$, see Figure 4.4. The important properties are that ψ_δ converges uniformly to the identity as $\delta \to 0$ and that $\psi_\delta(\mathrm{e}^{i\vartheta})$ converges *tangentially* to $\mathrm{e}^{i\vartheta}$ for certain points $\mathrm{e}^{i\vartheta} \in \partial\mathbb{D}$.

- The idea is to consider $f^* \circ \psi_\delta$, where f^* is the solution of (H-OPT$_\infty$), i.e., $\|f^* - \varphi\|_{L^\infty(K)} = \min_{f \in H^\infty(\mathbb{D}), |f| \leq g} \|f - \varphi\|_{L^\infty(K)} = \tau^*$. Importantly, $f^* \circ \psi_\delta \in \mathcal{A}(\mathbb{D})$. In Lemma 4.19 we prove that $\limsup_{\delta \to 0} \|f^* \circ \psi_\delta - \varphi\|_{L^\infty(K)} \leq \tau^*$. Tangential convergence of ψ_δ at certain points is a crucial ingredient of the proof.

- This does not prove Theorem 4.15 yet, since $f^* \circ \psi_\delta$ may not be feasible for ($\mathcal{A}$-OPT$_\infty$), i.e., we may not have $|f^* \circ \psi_\delta| \leq g$. However, we can multiply $f^* \circ \psi_\delta$ by some positive η such that $\eta(f^* \circ \psi_\delta)$ is feasible for ($\mathcal{A}$-OPT$_\infty$) and $\eta = \eta(\delta) \to 1$ as $\delta \to 0$. It will turn out that

$\|\eta(\delta)(f^* \circ \psi_\delta) - \varphi\|_{L^\infty(K)}$ converges to the minimum of (H-OPT$_\infty$) as $\delta \to 0$, which finishes the proof of Theorem 4.15.

In the following, we use the multivalued complex argument function that maps a complex number z with polar representation $z = re^{i\vartheta}$, $r > 0$, $\vartheta \in \mathbb{R}$, to the set $\vartheta + 2\pi\mathbb{Z}$. The advantage of using the multivalued argument is that rules like $\arg(zw) = \arg z + \arg w$, $z, w \in \mathbb{C} \setminus \{0\}$, hold, which are more tedious to write down if one restricts the argument, e.g., to $[-\pi, \pi)$. However, we do not make this explicit in our notation, i.e., we write $\arg z = \vartheta$ instead of $\arg z = \vartheta + 2\pi\mathbb{Z}$. We also write $\arg z \in I$ to express that there is some $\vartheta \in \arg z$ with $\vartheta \in I$.

After this sort of small talk we finally start with the proof of Theorem 4.15.

Lemma 4.16. *Let $p : \partial\mathbb{D} \to [0, \infty)$ be Lipschitz continuous and let*

$$h(z) = \frac{1}{2\pi} \int_0^{2\pi} \frac{e^{i\vartheta} + z}{e^{i\vartheta} - z} p(e^{i\vartheta})\, d\vartheta, \qquad z \in \mathbb{D}.$$

For $\delta \in (0, 1)$ let $F_\delta(z) = z(1 - \delta h(z))$ and let

$$\psi_\delta(z) = \frac{F_\delta(z)}{\|F_\delta\|_{H^\infty(\mathbb{D})}}(1 - \delta^2).$$

Then the following statements hold true.

(a) *$\psi_\delta \in \mathcal{A}(\mathbb{D})$ and $\psi_\delta(\overline{\mathbb{D}}) \subset \mathbb{D}$.*

(b) *ψ_δ converges uniformly to the identity as $\delta \to 0$. More precisely, there is a constant $C > 0$ such that $\max_{z \in \overline{\mathbb{D}}} |\psi_\delta(z) - z| \le C\delta$, $\delta \in (0, 1)$.*

(c) *If $\operatorname{Re} h(e^{i\vartheta}) = 0$ and $\operatorname{Im} h(e^{i\vartheta}) \ne 0$, then $\psi_\delta(e^{i\vartheta}) \to e^{i\vartheta}$ tangentially as $\delta \to 0$. More precisely,*

$$\arg \psi_\delta(e^{i\vartheta}) = \vartheta + \arctan\left(-\delta \operatorname{Im} h(e^{i\vartheta})\right),$$

and there are $\delta_0 > 0$ and $C > 0$ such that for $\delta \in (0, \delta_0)$ and for all $e^{i\vartheta}$ with $\operatorname{Re} h(e^{i\vartheta}) = 0$ and $\operatorname{Im} h(e^{i\vartheta}) \ne 0$

$$\left|1 - |\psi_\delta(e^{i\vartheta})|\right| \le C\delta^2.$$

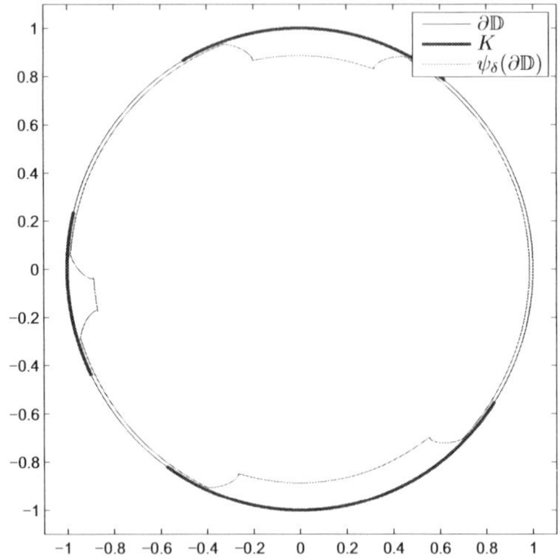

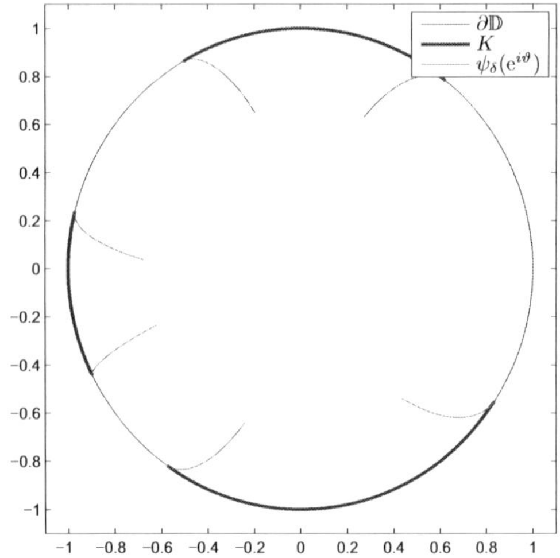

Figure 4.4: *Top*: The bold curves on the circle $\partial\mathbb{D}$ comprise some set K. The curve inside the circle is $\psi_\delta(\partial\mathbb{D})$ with $\delta = 0.1$. The function h that is needed in the construction of ψ_δ is 1 on some subset of K, 0 on $\partial\mathbb{D} \setminus K$ and linear in between. *Bottom*: The curce inside the circle show where ψ_δ maps the boundary points of K when δ varies between 0.5 and 0. The curces approach the boundary points of K tangentially to the unit circle and from the interior of the cone $\{\lambda K : \lambda \geq 0\}$.

Our definition of the function F_δ is inspired by a result of NEHARI [44, Chapter V.11] concerning conformal mapping from the unit disk to nearly circular domains. To the reader not too familiar with Hardy spaces we should point out that the real part of the function h is the Poisson integral of p, and especially $\operatorname{Re} h = p$ on $\partial\mathbb{D}$.

Proof. From the basic theory of Hardy spaces it is well-known that h is analytic on $\mathbb{D}$. Moreover, since p is Lipschitz continuous, $h \in \mathcal{A}(\mathbb{D})$, see, e.g., [25, Corollary III.1.4]. (Continuity of h actually follows already if p is only Dini continuous, but we are not going to use that.) Therefore $\psi_\delta \in \mathcal{A}(\mathbb{D})$. From the definition of ψ_δ it is clear that $\psi_\delta(\overline{\mathbb{D}}) \subset \mathbb{D}$. This is (a).

In order to see (b), notice first that $(1 - \delta|h(z)|)|z| \le |F_\delta(z)| \le (1 + \delta|h(z)|)|z|$ implies

$$|1 - \|F_\delta\|_{H^\infty(\mathbb{D})}| \le \delta \|h\|_{H^\infty(\mathbb{D})}.$$

Then

$$\begin{aligned}
|\psi_\delta(z) - z| &\le |F_\delta(z) - z| + \left| \frac{1-\delta^2}{\|F_\delta\|_{H^\infty(\mathbb{D})}} - 1 \right| |F_\delta(z)| \\
&\le \delta|zh(z)| + \left|1 - \delta^2 - \|F_\delta\|_{H^\infty(\mathbb{D})}\right| \frac{|F_\delta(z)|}{\|F_\delta\|_{H^\infty(\mathbb{D})}} \\
&\le \delta\|h\|_{H^\infty(\mathbb{D})} + \delta^2 + |1 - \|F_\delta\|_{H^\infty(\mathbb{D})}| \\
&\le (2\|h\|_{H^\infty(\mathbb{D})} + 1)\delta.
\end{aligned}$$

This is (b) with $C = 2\|h\|_{H^\infty(\mathbb{D})} + 1$.

It remains to prove (c). For any $\mathrm{e}^{i\vartheta} \in \partial\mathbb{D}$

$$\begin{aligned}
\arg \psi_\delta(\mathrm{e}^{i\vartheta}) &= \arg F_\delta(\mathrm{e}^{i\vartheta}) = \arg\left(\mathrm{e}^{i\vartheta}(1 - \delta h(\mathrm{e}^{i\vartheta}))\right) \\
&= \vartheta + \arg\left(1 - \delta h(\mathrm{e}^{i\vartheta})\right) = \vartheta + \arctan\left(\frac{-\delta \operatorname{Im} h(\mathrm{e}^{i\vartheta})}{1 - \delta \operatorname{Re} h(\mathrm{e}^{i\vartheta})}\right).
\end{aligned}$$

Let especially $\operatorname{Re} h(\mathrm{e}^{i\vartheta}) = 0$ and $\operatorname{Im} h(\mathrm{e}^{i\vartheta}) \ne 0$. Then

$$\left(\arg \psi_\delta(\mathrm{e}^{i\vartheta})\right) - \vartheta = \arctan\left(-\delta \operatorname{Im} h(\mathrm{e}^{i\vartheta})\right),$$

which converges linearly to zero as $\delta \to 0$. This is the first assertion of (c).

Since $\operatorname{Re} h$ is the Poisson integral of p and since $p \ge 0$ on $\partial\mathbb{D}$, we have $\operatorname{Re} h \ge 0$ on $\overline{\mathbb{D}}$, i.e., h only takes values in $\{\operatorname{Re} z \ge 0\}$. Then for $z \in \overline{\mathbb{D}}$ and

$\delta \leq \frac{2}{\|\operatorname{Re} h\|_{L^\infty(\partial\mathbb{D})}}$

$$\begin{aligned} |F_\delta(z)| &= |z||1-\delta h(z)| \leq \sqrt{|1-\delta \operatorname{Re} h(z)|^2 + \delta^2(\operatorname{Im} h(z))^2} \\ &\leq \sqrt{1+\delta^2(\operatorname{Im} h(z))^2} \leq 1 + \delta^2 \frac{(\operatorname{Im} h(z))^2}{2} \\ &\leq 1 + \delta^2 \frac{\|\operatorname{Im} h(z)\|^2_{L^\infty(\partial\mathbb{D})}}{2}. \end{aligned} \tag{4.7}$$

Both the condition for δ and the fact h only takes values in $\{\operatorname{Re} z \geq 0\}$ were needed for the second inequality. Moreover, if $\operatorname{Re} h(\mathrm{e}^{i\vartheta}) = 0$, then $|F_\delta(\mathrm{e}^{i\vartheta})| = \sqrt{1+\delta^2(\operatorname{Im} h(z))^2} \geq 1$, whence $\|F_\delta\|_{H^\infty(\mathbb{D})} \geq 1$. Therefore,

$$\begin{aligned} \big|1 - |\psi_\delta(\mathrm{e}^{i\vartheta})|\big| &= 1 - \frac{|F_\delta(\mathrm{e}^{i\vartheta})|}{\|F_\delta\|_{H^\infty(\mathbb{D})}}(1-\delta^2) \\ &= \frac{\|F_\delta\|_{H^\infty(\mathbb{D})} - (1-\delta^2)\sqrt{1+\delta^2(\operatorname{Im} h(\mathrm{e}^{i\vartheta}))^2}}{\|F_\delta\|_{H^\infty(\mathbb{D})}} \\ &\leq \|F_\delta\|_{H^\infty(\mathbb{D})} - (1-\delta^2) \\ &\overset{(4.7)}{\leq} \delta^2 \left(\frac{\|\operatorname{Im} h(z)\|^2_{L^\infty(\partial\mathbb{D})}}{2} + 1 \right), \end{aligned}$$

which converges quadratically to zero as $\delta \to 0$. The second assertion of (c) therefore holds true with $C = \frac{\|\operatorname{Im} h(z)\|^2_{L^\infty(\partial\mathbb{D})}}{2} + 1$. To summarize, we have proved that the argument of $\psi_\delta(\mathrm{e}^{i\vartheta})$ converges linearly as $\delta \to 0$, while its modulus converges quadratically. This means that $\psi_\delta(\mathrm{e}^{i\vartheta}) \to \mathrm{e}^{i\vartheta}$ tangentially. □

We are going to apply Lemma 4.16 to a certain function p which we construct in the following lemma.

Lemma 4.17. *There is a Lipschitz continuous function $p : \partial\mathbb{D} \to [0,\infty)$ such that*

$$h(z) = \frac{1}{2\pi} \int_0^{2\pi} \frac{\mathrm{e}^{i\vartheta}+z}{\mathrm{e}^{i\vartheta}-z} p(\mathrm{e}^{i\vartheta})\, \mathrm{d}\vartheta$$

satisfies $\operatorname{Im} h(\mathrm{e}^{i\lambda_j}) < 0$, $\operatorname{Im} h(\mathrm{e}^{i\rho_j}) > 0$ and $\operatorname{Re} h(\mathrm{e}^{i\vartheta}) = 0$ for $\mathrm{e}^{i\vartheta}$ in some neighborhood of the points $\mathrm{e}^{i\lambda_1}, \ldots, \mathrm{e}^{i\lambda_n}$ and $\mathrm{e}^{i\rho_1}, \ldots, \mathrm{e}^{i\rho_n}$.

Proof. We begin with some simple estimates. First of all, recall from basic theory of Hardy spaces that for Lipschitz continuous p

$$\operatorname{Im} h(\mathrm{e}^{i\vartheta}) = \frac{1}{2\pi}\int_{-\pi}^{\pi} p(\mathrm{e}^{i(\vartheta-t)})\cot(t/2)\,\mathrm{d}t,$$

where the integral exists as a principal value integral (see, e.g., [32, Chapter 6]). Now assume that $p : \partial\mathbb{D} \to [0,\infty)$ is some Lipschitz continuous function with $0 \le p \le 1$, $\operatorname{supp} p \subset \mathrm{e}^{i[0,\sigma]}$ for some $0 < \sigma \le \pi$, and $p(\mathrm{e}^{i\vartheta}) = 1$ for $\vartheta \in [\epsilon, \sigma-\epsilon]$ for some small $\epsilon > 0$. Then we have the estimate

$$\begin{aligned}\operatorname{Im} h(\mathrm{e}^{i0}) &= \frac{1}{2\pi}\int_{-\pi}^{\pi} p(\mathrm{e}^{-it})\cot(t/2)\,\mathrm{d}t = \frac{1}{2\pi}\int_{-\sigma}^{0} p(\mathrm{e}^{-it})\cot(t/2)\,\mathrm{d}t \\ &\le \frac{1}{2\pi}\int_{-\sigma+\epsilon}^{-\epsilon} \cot(t/2)\,\mathrm{d}t = -\frac{1}{\pi}\ln\left(\frac{\sin((\sigma-\epsilon)/2)}{\sin(\epsilon/2)}\right).\end{aligned} \tag{4.8}$$

Notice that we used $\sigma \le \pi$ for the inequality so that $\cot(t/2) \le 0$ for $t \in [-\sigma, 0]$. Similarly,

$$\operatorname{Im} h(\mathrm{e}^{i\sigma}) \ge \frac{1}{\pi}\ln\left(\frac{\sin((\sigma-\epsilon)/2)}{\sin(\epsilon/2)}\right). \tag{4.9}$$

Moreover, if $\mathrm{e}^{i\vartheta} \in \partial\mathbb{D}$ such that $p = 0$ on $\mathrm{e}^{i[\vartheta-\eta,\vartheta+\eta]}$ for some $\eta > 0$, then

$$\begin{aligned}|\operatorname{Im} h(\mathrm{e}^{i\vartheta})| &= \frac{1}{2\pi}\left|\int_{-\pi}^{\pi} p(\mathrm{e}^{i(\vartheta-t)})\cot(t/2)\,\mathrm{d}t\right| \\ &\le \frac{1}{\pi}\int_{\eta}^{\pi} \cot(t/2)\,\mathrm{d}t = -\frac{2}{\pi}\ln(\sin(\eta/2)).\end{aligned} \tag{4.10}$$

We are now going to construct a function p that satisfies all of the assertions of the lemma. Without loss of generality we can assume that $|K_j| \le \pi$ for all j. If this is not true, we can apply a Möbius transformation. Let $d_0 = \min_{j\ne l}\operatorname{dist}(K_j, K_l)$ and $M = -\frac{2}{\pi}\ln(\sin(d_0/2))$. Let $\epsilon > 0$ be so small that for all $j \in \{1,\dots,n\}$

$$\frac{1}{\pi}\ln\left(\frac{\sin((|K_j|-\epsilon)/2)}{\sin(\epsilon/2)}\right) \ge nM. \tag{4.11}$$

For each j let p_j be a Lipschitz continuous function on $\partial\mathbb{D}$ such that $\operatorname{supp} p_j \subset K_j = \mathrm{e}^{i[\lambda_j,\rho_j]}$, $0 \le p_j \le 1$ and $p_j = 1$ on $\mathrm{e}^{i[\lambda_j+\epsilon,\rho_j-\epsilon]}$. Then $p = \sum_{j=1}^{n} p_j$ satisfies all of the assertions of the lemma. Indeed, let

$$h_j(z) = \frac{1}{2\pi}\int_0^{2\pi} \frac{\mathrm{e}^{i\vartheta}+z}{\mathrm{e}^{i\vartheta}-z} p_j(\mathrm{e}^{i\vartheta})\,\mathrm{d}\vartheta$$

By (4.10) we have for every $j \in \{1, \dots, n\}$

$$\left|\sum_{l \neq j} \operatorname{Im} h_l(\mathrm{e}^{i\lambda_j})\right| \leq (n-1)M \qquad \text{and} \qquad \left|\sum_{l \neq j} \operatorname{Im} h_l(\mathrm{e}^{i\rho_j})\right| \leq (n-1)M, \tag{4.12}$$

and by (4.8), (4.9) and (4.11) we have

$$\operatorname{Im} h_j(\mathrm{e}^{i\lambda_j}) \leq -nM \qquad \text{and} \qquad \operatorname{Im} h_j(\mathrm{e}^{i\rho_j}) \geq nM. \tag{4.13}$$

(4.12) and (4.13) together give

$$\operatorname{Im} h(\mathrm{e}^{i\lambda_j}) \leq -M \qquad \text{and} \qquad \operatorname{Im} h(\mathrm{e}^{i\rho_j}) \geq M.$$

Concerning the statement about the real part of h, we use that we have some freedom left in the construction. We can choose the p_j such that $p_j = 0$ for all j in some small neighborhood of the points $\mathrm{e}^{i\lambda_1}, \dots, \mathrm{e}^{i\lambda_n}$ and $\mathrm{e}^{i\rho_1}, \dots, \mathrm{e}^{i\rho_n}$. The statement then follows from the fact that $\operatorname{Re} h(\mathrm{e}^{i\vartheta}) = p(\mathrm{e}^{i\vartheta})$ for all $\mathrm{e}^{i\vartheta} \in \partial\mathbb{D}$. □

The following lemma seems quite obvious to us, but we are not aware of any reference. We therefore prove it for the convenience of the reader. Recall that for $f \in L^\infty(\partial\mathbb{D})$ the *essential range* of f on a measurable set $I \subset \partial\mathbb{D}$ is

$$\operatorname{ess\,ran}(f, I) = \left\{ z \in \mathbb{C} : \begin{array}{l} f^{-1}(B_\epsilon(z)) \cap I \text{ has positive Lebesgue} \\ \text{measure for all } \epsilon > 0 \end{array} \right\}.$$

Here, $B_\epsilon(z) = \{w \in \mathbb{C} : |z - w| < \epsilon\}$ denotes a ball in $\mathbb{C}$. For example, if f is continuous and I is a closed interval, then the essential range conincides with the classical range, i.e., $\operatorname{ess\,ran}(f, I) = f(I) = \{f(\mathrm{e}^{i\vartheta}) : \mathrm{e}^{i\vartheta} \in I\}$. However, it is clear that the last expression does not make any sense for functions from $L^\infty(\partial\mathbb{D})$. Further, when $\mathrm{e}^{i[\vartheta_1, \vartheta_2]}$ is an interval on $\partial\mathbb{D}$, we also write $\operatorname{ess\,ran}(f, [\vartheta_1, \vartheta_2])$ instead of $\operatorname{ess\,ran}(f, \mathrm{e}^{i[\vartheta_1, \vartheta_2]})$ for simplicity of notation.

Lemma 4.18. *Let $f \in H^\infty(\mathbb{D})$. Then for any $\epsilon > 0$ there is $\delta > 0$ such that if $z \in \mathbb{D}$ and $\mathrm{e}^{i\vartheta} \in \partial\mathbb{D}$ with $|z - \mathrm{e}^{i\vartheta}| \leq \delta$, then*

$$f(z) \in \operatorname{conv}(\operatorname{ess\,ran}(f, [\vartheta - \epsilon, \vartheta + \epsilon])) + B_\epsilon(0).$$

Proof. Fix some $\epsilon > 0$. Let $\delta > 0$ such that $\arg \mathrm{e}^{iB_\delta(0)} \subset [-\frac{\epsilon}{2}, \frac{\epsilon}{2}]$ and such that for all $r \in (1 - \delta, 1)$

$$\left| 1 - \frac{1}{\frac{1}{2\pi} \int_{[-\epsilon/2, \epsilon/2]} P_r(t)\, \mathrm{d}t} \right| \leq \frac{\epsilon}{2\|f\|_{H^\infty\mathbb{D})}} \tag{4.14}$$

and

$$\frac{1}{2\pi}\int_{[-\pi,\pi]\setminus[-\epsilon/2,\epsilon/2]} P_r(t)\,\mathrm{d}t \le \frac{\epsilon}{2\|f\|_{H^\infty\mathbb{D})}}. \tag{4.15}$$

This is possible, because the Poisson kernel is an approximate identity.

Now let $z \in \mathbb{D}$ and $\mathrm{e}^{i\vartheta} \in \partial\mathbb{D}$ with $|z - \mathrm{e}^{i\vartheta}| \le \delta$ and write $z = r\mathrm{e}^{i\tau}$ with $r \ge 0$ and $\tau \in \mathbb{R}$. Notice that $r \in (1-\delta, 1)$. Then

$$f(z) = \frac{1}{2\pi}\left(\int_{[-\epsilon/2,\epsilon/2]} + \int_{[-\pi,\pi]\setminus[-\epsilon/2,\epsilon/2]} f(\mathrm{e}^{i(\tau-t)})P_r(t)\,\mathrm{d}t\right) =: I_1 + I_2. \tag{4.16}$$

For the second integral it follows from (4.15) that

$$|I_2| \le \frac{\epsilon}{2}. \tag{4.17}$$

For the first integral we have

$$\begin{aligned} I_1 &= \left(\frac{1}{\frac{1}{2\pi}\int_{[-\epsilon/2,\epsilon/2]} P_r(t)\,\mathrm{d}t}\right)\frac{1}{2\pi}\int_{[-\epsilon/2,\epsilon/2]} f(\mathrm{e}^{i(\tau-t)})P_r(t)\,\mathrm{d}t \\ &\quad + \left(1 - \frac{1}{\frac{1}{2\pi}\int_{[-\epsilon/2,\epsilon/2]} P_r(t)\,\mathrm{d}t}\right)\frac{1}{2\pi}\int_{[-\epsilon/2,\epsilon/2]} f(\mathrm{e}^{i(\tau-t)})P_r(t)\,\mathrm{d}t \\ &=: I_1' + I_1''. \end{aligned} \tag{4.18}$$

From (4.14) we get

$$|I_1''| \le \frac{\epsilon}{2}. \tag{4.19}$$

Since $P_r \ge 0$ and $\left(\frac{1}{\frac{1}{2\pi}\int_{[-\epsilon/2,\epsilon/2]} P_r(t)\,\mathrm{d}t}\right)\frac{1}{2\pi}\int_{[-\epsilon/2,\epsilon/2]} P_r(t)\,\mathrm{d}t = 1$, it follows that

$$I_1' \in \operatorname{conv}(\operatorname{ess\,ran}(f, [\tau - \epsilon/2, \tau + \epsilon/2])).$$

From $\arg \mathrm{e}^{iB_\delta(0)} \subset [-\frac{\epsilon}{2}, \frac{\epsilon}{2}]$ it follows that $\tau = \arg z \in [\vartheta - \frac{\epsilon}{2}, \vartheta + \frac{\epsilon}{2}]$, whence $[\tau - \epsilon/2, \tau + \epsilon/2] \subset [\vartheta - \epsilon, \vartheta + \epsilon]$. Therefore,

$$I_1' \in \operatorname{conv}(\operatorname{ess\,ran}(f, [\vartheta - \epsilon, \vartheta + \epsilon])). \tag{4.20}$$

Now (4.16)–(4.20) together yield

$$f(z) = I_1' + I_1'' + I_2 \in \operatorname{conv}(\operatorname{ess\,ran}(f, [\vartheta - \epsilon, \vartheta + \epsilon])) + B_\epsilon(0). \qquad \square$$

A big step towards the proof of Theorem 4.15 is

Lemma 4.19. *Let h be a function with the properties from Lemma 4.17. For $\delta > 0$ let ψ_δ be constructed from h as in Lemma 4.16. Let f^* be the solution of* (H-OPT$_\infty$) *and $\tau^* = \|f^* - \varphi\|_{L^\infty(K)}$. Then*

$$\limsup_{\delta \to 0} \|f^* \circ \psi_\delta - \varphi\|_{L^\infty(K)} \leq \tau^*.$$

Proof. Fix $\epsilon > 0$. Write $h = u + iv$. Let $\delta_0 > 0$ be so small that

$$|t| \leq 2\delta_0 \|v\|_{L^\infty(\partial\mathbb{D})} \quad \Rightarrow \quad |\varphi(\mathrm{e}^{i(\vartheta+t)}) - \varphi(\mathrm{e}^{i\vartheta})| \leq \frac{\epsilon}{2}. \tag{4.21}$$

By the properties of h from Lemma 4.17 there is $\eta > 0$ such that for all $\vartheta \in \bigcup_{j=1}^n [\lambda_j, \lambda_j + \eta]$ we have $u(\mathrm{e}^{i\vartheta}) = 0$ and $v(\mathrm{e}^{i\vartheta}) \leq m < 0$ for some m. Since v is bounded away from zero, there is a constant $C_1 > 0$ such that for $0 < \delta \leq \delta_0$ and all $\vartheta \in \bigcup_{j=1}^n [\lambda_j, \lambda_j + \eta]$

$$C_1 \delta \leq \arctan(-\delta v(\mathrm{e}^{i\vartheta})) \leq \|v\|_{L^\infty(\partial\mathbb{D})} \delta. \tag{4.22}$$

Now let $\vartheta \in [\lambda_j, \lambda_j + \eta]$ for some j and $0 < \delta \leq \delta_0$. Write $\psi_\delta(\mathrm{e}^{i\vartheta}) = r\mathrm{e}^{i\tau}$ with $r \geq 0$ and real τ. Then

$$\begin{aligned} &|(f^* \circ \psi_\delta)(\mathrm{e}^{i\vartheta}) - \varphi(\mathrm{e}^{i\vartheta})| \\ &\quad = |f^*(r\mathrm{e}^{i\tau}) - \varphi(\mathrm{e}^{i\vartheta})| = \frac{1}{2\pi} \left| \int_{-\pi}^{\pi} (f^*(\mathrm{e}^{i(\tau-t)}) - \varphi(\mathrm{e}^{i\vartheta})) P_r(t)\, \mathrm{d}t \right| \\ &\quad \leq \frac{1}{2\pi} \left(\int_{[-C_1\delta, C_1\delta]} + \int_{[-\pi,\pi]\setminus[-C_1\delta, C_1\delta]} |f^*(\mathrm{e}^{i(\tau-t)}) - \varphi(\mathrm{e}^{i\vartheta})| P_r(t)\, \mathrm{d}t \right). \end{aligned} \tag{4.23}$$

We estimate the first integral. Let $|t| \leq \delta C_1$. From Lemma 4.16(c) we have

$$(\tau - t) - \vartheta = \arctan(-\delta v(\mathrm{e}^{i\vartheta})) - t \pmod{2\pi}. \tag{4.24}$$

Further,

$$|\arctan(-\delta v(\mathrm{e}^{i\vartheta})) - t| \overset{(4.22)}{\leq} (\|v\|_{L^\infty(\partial\mathbb{D})} + C_1)\delta \leq 2\delta_0 \|v\|_{L^\infty(\partial\mathbb{D})}.$$

Now (4.21) implies $|\varphi(\mathrm{e}^{i(\tau-t)}) - \varphi(\mathrm{e}^{i\vartheta})| \leq \frac{\epsilon}{2}$, so

$$\begin{aligned} |f^*(\mathrm{e}^{i(\tau-t)}) - \varphi(\mathrm{e}^{i\vartheta})| &\leq |f^*(\mathrm{e}^{i(\tau-t)}) - \varphi(\mathrm{e}^{i(\tau-t)})| + |\varphi(\mathrm{e}^{i(\tau-t)}) - \varphi(\mathrm{e}^{i\vartheta})| \\ &\leq |f^*(\mathrm{e}^{i(\tau-t)}) - \varphi(\mathrm{e}^{i(\tau-t)})| + \frac{\epsilon}{2}. \end{aligned} \tag{4.25}$$

We now show that $e^{i(\tau-t)} \in K_j = e^{i[\lambda_j,\rho_j]}$ for δ small enough. From (4.24) we have $\tau - t = \vartheta + \arctan(-\delta v(e^{i\vartheta})) - t \pmod{2\pi}$. Now let $\delta_1 > 0$ be so small that $\eta + (C_1 + \|v\|_{L^\infty(\partial\mathbb{D})})\delta_1 \leq |K_j|$ for all j and assume further that $0 < \delta \leq \delta_1$. Then by choice of ϑ, (4.22) and choice of t

$$\begin{aligned}\vartheta + \arctan(-\delta v(e^{i\vartheta})) - t &\in [\lambda_j, \lambda_j + \eta] + [C_1\delta, \|v\|_{L^\infty(\partial\mathbb{D})}\delta] + [-C_1\delta, C_1\delta] \\ &= [\lambda_j, \lambda_j + \eta + (C_1 + \|v\|_{L^\infty(\partial\mathbb{D})})\delta] \\ &\subset [\lambda_j, \rho_j].\end{aligned}$$

It follows that $e^{i(\tau-t)} \in K_j$. By assumption, $\|f^* - \varphi\|_{L^\infty(K)} \leq \tau^*$, so

$$|f^*(e^{i(\tau-t)}) - \varphi(e^{i(\tau-t)})| \leq \tau^*. \tag{4.26}$$

(4.25) and (4.26) together give

$$|f^*(e^{i(\tau-t)}) - \varphi(e^{i\vartheta})| \leq \tau^* + \frac{\epsilon}{2},$$

and therefore the first integral in (4.23) can be estimated by

$$\frac{1}{2\pi}\int_{[-C_1\delta,C_1\delta]} |f^*(e^{i(\tau-t)}) - \varphi(e^{i\vartheta})| P_r(t)\,dt \leq \tau^* + \epsilon/2. \tag{4.27}$$

We estimate the second integral in (4.23) by

$$\begin{aligned}&\frac{1}{2\pi}\int_{[-\pi,\pi]\setminus[-C_1\delta,C_1\delta]} |f^*(e^{i(\tau-t)}) - \varphi(e^{i\vartheta})| P_r(t)\,dt \\ &\quad\leq \left(\int_{[-\pi,\pi]\setminus[-C_1\delta,C_1\delta]} P_r(t)\,dt\right)\left(\frac{1}{\pi}\|g\|_{L^\infty(\partial\mathbb{D})}\right).\end{aligned} \tag{4.28}$$

A straightforward calculation shows that

$$\begin{aligned}\int_{[-\pi,\pi]\setminus[-C_1\delta,C_1\delta]} P_r(t)\,dt &= 2\pi - 4\arctan\left(\frac{1+r}{1-r}\tan\left(\frac{C_1\delta}{2}\right)\right) \\ &\leq 2\pi - 4\arctan\left(\frac{1}{C_2\delta^2}\tan\left(\frac{C_1\delta}{2}\right)\right),\end{aligned}$$

where we recall that $r = |\psi_\delta(e^{i\vartheta})|$ and C_2 is the constant from Lemma 4.16(c). The last expression converges to zero as $\delta \to 0$. We want to emphasize that this is the point where tangential convergence is needed: In order for the expression inside of the arctan to converge to infinity, it is necessary that

$1-r=1-|\psi_\delta(\mathrm{e}^{i\vartheta})|$ converges faster to zero than linearly in δ. We conclude that there is $\delta_2>0$ such that $0<\delta\leq\delta_2$ implies that the expression on the right hand side of (4.28) is smaller than $\frac{\epsilon}{2}$. Combining this with the estimates (4.23) and (4.27) we obtain that if $0<\delta\leq\min\{\delta_0,\delta_1,\delta_2\}$, then for all $\vartheta\in\bigcup_{j=1}^n[\lambda_j,\lambda_j+\eta]$

$$|(f^*\circ\psi_\delta)(\mathrm{e}^{i\vartheta})-\varphi(\mathrm{e}^{i\vartheta})|\leq\tau^*+\epsilon. \tag{4.29}$$

Similarly, one can show that there is $\delta_3>0$ so that this inequality holds for $0<\delta\leq\delta_3$ and all $\vartheta\in\bigcup_{j=1}^n[\rho_j-\widetilde{\eta},\rho_j]$ with some $\widetilde{\eta}>0$.

It remains to show that for small enough δ the inequality holds for $\vartheta\in\bigcup_{j=1}^n[\lambda_j+\eta,\rho_j-\widetilde{\eta}]$. This is an easy consequence of Lemma 4.18. By uniform continuity there is $\epsilon_1>0$ such that $\epsilon_1\leq\frac{\epsilon}{2}$, $\epsilon_1\leq\max\{\eta,\widetilde{\eta}\}$ and such that

$$|t|\leq\epsilon_1 \quad\Rightarrow\quad |\varphi(\mathrm{e}^{i(\vartheta+t)})-\varphi(\mathrm{e}^{i\vartheta})|\leq\frac{\epsilon}{2}. \tag{4.30}$$

By Lemma 4.18 there is $\epsilon_2>0$ such that

$$|\mathrm{e}^{i\vartheta}-z|<\epsilon_2 \quad\Rightarrow\quad f^*(z)\in\operatorname{conv}(\operatorname{ess\,ran}(f^*,[\vartheta-\epsilon_1,\vartheta+\epsilon_1]))+B_{\epsilon_1}(0). \tag{4.31}$$

Finally, by Lemma 4.16(b) there is $\delta_4>0$ such that for all $0<\delta\leq\delta_4$ we have $\max_{z\in\mathbb{D}}|\psi_\delta(z)-z|\leq\epsilon_2$. Now let $\vartheta\in[\lambda_j+\eta,\rho_j-\widetilde{\eta}]$ for some j. Then for $0<\delta\leq\delta_4$ we have $|\psi_\delta(\mathrm{e}^{i\vartheta})-\mathrm{e}^{i\vartheta}|\leq\epsilon_2$, so

$$\begin{aligned}
f^*(\psi_\delta(\mathrm{e}^{i\vartheta})) &\overset{(4.31)}{\in} \operatorname{conv}(\operatorname{ess\,ran}(f^*,[\vartheta-\epsilon_1,\vartheta+\epsilon_1]))+B_{\epsilon_1}(0)\\
&\subset \operatorname{conv}\left(\bigcup_{|t|\leq\epsilon_1}B_{\tau^*}(\varphi(\mathrm{e}^{i(\vartheta+t)}))\right)+B_{\epsilon/2}(0)\\
&\qquad\qquad\text{since } \epsilon_1\leq\max\{\eta,\widetilde{\eta}\}\\
&\overset{(4.30)}{\subset} B_{\tau^*+\epsilon/2}(\varphi(\mathrm{e}^{i\vartheta}))+B_{\epsilon/2}(0)=B_{\tau^*+\epsilon}(\varphi(\mathrm{e}^{i\vartheta})).
\end{aligned}$$

This is just equation (4.29).

Summing up, we have shown that if $0<\delta\leq\min\{\delta_0,\dots,\delta_4\}$, then $\|f^*\circ\psi_\delta-\varphi\|_{L^\infty(K)}\leq\tau^*+\epsilon$. Because $\epsilon>0$ was arbitrary, this proves the lemma. □

Using the work we have done so far it is not hard any more to prove Theorem 4.15.

Proof of Theorem 4.15. Let f^* be the solution of (H-OPT$_\infty$) and $\tau^* = \|f^* - \varphi\|_{L^\infty(K)}$. Fix $\epsilon > 0$. Let $\epsilon_1 > 0$ such that with

$$\eta = \left(1 + \frac{\epsilon_1}{\min_{e^{i\vartheta} \in \partial\mathbb{D}} g(e^{i\vartheta})}\right)^{-1}$$

we have $(1-\eta)\|f^*\|_{H^\infty(\mathbb{D})} < \epsilon/2$. Because ψ_δ converges uniformly to the identity as $\delta \to 0$ and since g is uniformly continuous, it follows from Lemma 4.18 as in the proof of Lemma 4.19 that for $\delta > 0$ small enough

$$|(f^* \circ \psi_\delta)(e^{i\vartheta})| \leq g(e^{i\vartheta}) + \epsilon_1.$$

By Lemma 4.19 we have for $\delta > 0$ small enough

$$\|f^* \circ \psi_\delta - \varphi\|_{L^\infty(K)} \leq \tau^* + \frac{\epsilon}{2}.$$

From Lemma 4.16(a) it follows that $f^* \circ \psi_\delta \in \mathcal{A}(\mathbb{D})$. Moreover, for $e^{i\vartheta} \in \partial\mathbb{D}$

$$|(f^* \circ \psi_\delta)(e^{i\vartheta})| \leq g(e^{i\vartheta}) + \epsilon_1 \leq \left(1 + \frac{\epsilon_1}{\min_{e^{i\tau} \in \partial\mathbb{D}} g(e^{i\tau})}\right) g(e^{i\vartheta}),$$

whence $|\eta(f^* \circ \psi_\delta)| \leq g$. This means that $\eta(f^* \circ \psi_\delta)$ is feasible for ($\mathcal{A}$-OPT$_\infty$). Then

$$\begin{aligned}
\tau^* &= \min_{f \in H^\infty(\mathbb{D}), |f| \leq g} \|f - \varphi\|_{L^\infty(K)} \leq \inf_{f \in \mathcal{A}(\mathbb{D}), |f| \leq g} \|f - \varphi\|_{L^\infty(K)} \\
&\leq \|\eta(f^* \circ \psi_\delta) - \varphi\|_{L^\infty(K)} \leq \|f^* \circ \psi_\delta - \varphi\|_{L^\infty(K)} + (1-\eta)\|f^*\|_{H^\infty(\mathbb{D})} \\
&\leq \tau^* + \frac{\epsilon}{2} + \frac{\epsilon}{2} = \tau^* + \epsilon.
\end{aligned}$$

Since $\epsilon > 0$ was arbitrary, $\min_{f \in H^\infty(\mathbb{D}), |f| \leq g} \|f - \varphi\|_{L^\infty(K)} = \inf_{f \in \mathcal{A}(\mathbb{D}), |f| \leq g} \|f - \varphi\|_{L^\infty(K)}$. This is what we had to prove. □

Chapter 5

Constrained Optimization in Hardy Spaces: Numerics

MOSEK seems to crash. Is that a feature?

THE MOSEK FREQUENTLY ASKED QUESTIONS (V. 5.0)

We want to solve (H-OPT$_p$) (or ($\mathcal{A}$-OPT$_p$)) from Chapter 4 numerically. In Section 5.1 we devise a general discretization scheme for (H-OPT$_p$) and show that the minimum of the discrete problem converges to the minimum (or infimum) of the continuous problem as the discretization becomes better. Moreover, we can even show convergence of the minimizing functions. In Section 5.2 we consider several concrete discretizations. In the cases $p = 2$ and $p = \infty$ we obtain *quadratically constrained quadratic programs (QCQPs)*, which we write down more explicitly in Section 5.3. In order to solve these problems, we reformulate them as *second-order cone programs (SOCPs)*. This class of problems is briefly introduced in Section 5.4, and the SOCP formulations of our problems are derived in Section 5.5. We finish with numerical examples in Section 5.6.

5.1 Discretization

We discretize (H-OPT$_p$) in two steps. First, we discretize the space $H^\infty(\mathbb{D})$ and obtain a semi-discrete problem. We show that the minimum of the semi-discrete problem converges to the minimum (or infimum) of the continuous problem as the dimension of the discrete space tends to infinity (Theorem 5.1). Moreover, we show convergence of the minimizing functions (Corollary 5.3), in the cases $p = 1$ and $p = \infty$ under the additional assumptions of Theorem 4.4 and Theorem 4.15, respectively. Next, we obtain a fully discrete problem that can be solved on a computer by checking the constraint $|f| \leq g$ on $\partial\mathbb{D}$ only on a grid and replacing the integral from the objective function by a quadrature approximation. We show that the minimum of the fully discrete problem converges to the minimum of the semi-discrete problem as the grid becomes finer and the quadrature approximation becomes better (Theorem 5.4). The culmination of this section is Theorem 5.10, which states that the discretization parameters can be chosen in such a way that the minimum of the fully discrete problem converges to the minimum of (H-OPT$_p$) as the discretization becomes better, in the cases $p = 1$ and $p = \infty$ again under the additional assumptions of Theorem 4.4 and Theorem 4.15, respectively. Moreover, we have convergence of the minimizing functions.

We begin with some notation.

5.1.1 Assumptions and notation

From now on we assume $K = \overline{K}$ and that φ and g are real symmetric so that the solution of (H-OPT$_p$) is also real symmetric (Theorem 4.11). Optimization therefore takes place in the space

$$\begin{aligned}\mathcal{H}^\infty(\mathbb{D}) &= \left\{ f \in H^\infty(\mathbb{D}) : f(\mathrm{e}^{i\vartheta}) = \overline{f(\mathrm{e}^{-i\vartheta})},\ \mathrm{e}^{i\vartheta} \in \partial\mathbb{D} \right\} \\ &= \left\{ f \in L^\infty(\partial\mathbb{D}) : \begin{array}{l} \widehat{f}_k = 0 \text{ for integers } k < 0, \\ \widehat{f}_k \in \mathbb{R} \text{ for integers } k \geq 0 \end{array} \right\}\end{aligned}$$

instead of $H^\infty(\mathbb{D})$. (Also compare Theorem 2.5.)

Let $N \in \mathbb{N}$. For $\alpha = (\alpha_0, \alpha_1, \ldots, \alpha_{N-1})^\top \in \mathbb{R}^N$ write

$$f_\alpha(\mathrm{e}^{i\vartheta}) = \sum_{k=0}^{N-1} \alpha_k \mathrm{e}^{ik\vartheta}.$$

Moreover, for $\beta = (\beta_{-N}, \beta_{-N+1}, \ldots, \beta_{N-1})^\top \in \mathbb{R}^{2N}$ let

$$f_\beta(\mathrm{e}^{i\vartheta}) = \sum_{k=-N}^{N-1} \beta_k \mathrm{e}^{ik\vartheta}.$$

For computations we are going to use the finite dimensional subspaces

$$\mathcal{H}^\infty_N(\mathbb{D}) = \left\{ f_\alpha : f_\alpha(\mathrm{e}^{i\vartheta}) = \sum_{k=0}^{N-1} \alpha_k \mathrm{e}^{ik\vartheta},\ \alpha = (\alpha_0, \ldots, \alpha_{N-1})^\top \in \mathbb{R}^N \right\}$$

and

$$\mathcal{L}_N^\infty(\partial\mathbb{D}) = \left\{ f_\beta : f_\beta(\mathrm{e}^{i\vartheta}) = \sum_{k=-N}^{N-1} \beta_k \mathrm{e}^{ik\vartheta},\ \beta = (\beta_{-N}, \ldots, \beta_{N-1})^\top \in \mathbb{R}^{2N} \right\}.$$

By $\mathcal{X}$ we usually denote a grid on $\partial\mathbb{D}$, i.e., a set of finitely many points from $\partial\mathbb{D}$. Given two points $\mathrm{e}^{i\vartheta}$ and $\mathrm{e}^{i\tau}$ on $\partial\mathbb{D}$, we define the *distance between* $\mathrm{e}^{i\vartheta}$ *and* $\mathrm{e}^{i\tau}$ to be

$$\operatorname{dist}\left(\mathrm{e}^{i\vartheta}, \mathrm{e}^{i\tau}\right) = \min_{\mathrm{e}^{i(\vartheta+\mu)} = \mathrm{e}^{i\tau}} |\mu|.$$

Clearly, we always have $\operatorname{dist}\left(\mathrm{e}^{i\vartheta}, \mathrm{e}^{i\tau}\right) \leq \pi$. The *fineness of the grid* $\mathcal{X}$, i.e., the maximal distance between two neighboring points, is

$$h_{\max}(\mathcal{X}) = \max_{\mathrm{e}^{i\vartheta} \in \mathcal{X}} \min_{\mathrm{e}^{i\tau} \in \mathcal{X} \setminus \{\mathrm{e}^{i\vartheta}\}} \operatorname{dist}\left(\mathrm{e}^{i\vartheta}, \mathrm{e}^{i\tau}\right).$$

5.1.2 Semi-discrete problem

The first step to obtain a discretization is to replace the space $H^\infty(\mathbb{D})$ in (H-OPT$_p$) by the discrete space $\mathcal{H}_N^\infty(\mathbb{D})$. (Recall that we only look for real symmetric solutions.) We therefore consider the semi-discrete problem

$$\begin{array}{ll} \text{minimize} & \|f - \varphi\|_{L^p(K)} \\ \text{subject to} & |f| \leq g \quad \text{on } \partial\mathbb{D}, \\ & f \in \mathcal{H}_N^\infty(\mathbb{D}). \end{array} \tag{SDP$_p$}$$

We also write, e.g., (SDP$_p(N)$) or (SDP$_p(N,\varphi)$) in order to denote the above problem with a specific N or a specific φ. For $1 < p < \infty$, (SDP$_p$) has a unique solution since the objective function is strictly convex and we are minimizing over a compact and convex set. In the cases $p = 1$ and $p = \infty$, the solution may not be unique since the objective function is convex, but not strictly convex.

Theorem 5.1. *If* $1 \leq p < \infty$, *then the minimum of* (SDP$_p(N)$) *converges to the minimum of* (H-OPT$_p$) *as* $N \to \infty$, *that is, if* f_N^* *is a solution of* (SDP$_p(N)$), *and* f^* *is a solution of* (H-OPT$_p$), *then*

$$\|f_N^* - \varphi\|_{L^p(K)} \to \|f^* - \varphi\|_{L^p(K)} \qquad \text{as } N \to \infty.$$

If $p = \infty$, *then the minimum of* (SDP$_\infty(N)$) *converges to the infimum of* ($\mathcal{A}$-OPT$_\infty$) *as* $N \to \infty$.

Proof. We first consider the case $1 \le p < \infty$. Fix $\epsilon > 0$. By Theorem 4.13 there is a polynomial $\widetilde{f}$ with $|\widetilde{f}| \le g$ such that

$$\|\widetilde{f} - \varphi\|_{L^p(K)} \le \min_{f \in H^\infty(\mathbb{D}),\, |f| \le g} \|f - \varphi\|_{L^p(K)} + \epsilon.$$

Then for $N \ge \deg \widetilde{f} - 1$

$$\begin{aligned} \|f_N^* - \varphi\|_{L^p(K)} &= \min_{f \in \mathcal{H}_N^\infty(\mathbb{D}),\, |f| \le g} \|f - \varphi\|_{L^p(K)} \\ &\le \|\widetilde{f} - \varphi\|_{L^p(K)} \\ &\le \min_{f \in H^\infty(\mathbb{D}),\, |f| \le g} \|f - \varphi\|_{L^p(K)} + \epsilon \\ &= \|f^* - \varphi\|_{L^p(K)} + \epsilon. \end{aligned}$$

On the other hand, since f_N^* is feasible for (H-OPT$_p$),

$$\|f^* - \varphi\|_{L^p(K)} \le \|f_N^* - \varphi\|_{L^p(K)}.$$

Together we have for $N \ge \deg \widetilde{f} - 1$

$$\|f^* - \varphi\|_{L^p(K)} \le \|f_N^* - \varphi\|_{L^p(K)} \le \|f^* - \varphi\|_{L^p(K)} + \epsilon.$$

Therefore, $\|f_N^* - \varphi\|_{L^p(K)} \to \|f^* - \varphi\|_{L^p(K)}$ as $N \to \infty$.

The case $p = \infty$ can be handled in the same way by using second part of Theorem 4.13 instead. □

Remark 5.2. *If $g \in L^\infty(\partial\mathbb{D})$ and $\inf g > 0$, then the minimum of* (SDP$_p(N)$) *converges to the infimum of* ($\mathcal{A}$-OPT$_p$) *as $N \to \infty$. In order to see this, use Theorem 4.14 instead of Theorem 4.13 and adapt the preceding proof.*

Corollary 5.3. *Let f_N^* be a solution of* (SDP$_p(N)$)*, and f^* a solution of* (H-OPT$_p$)*. If $1 < p < \infty$, then (f_N^*) converges to f^* strongly in $L^p(\partial\mathbb{D})$. If $p = \infty$ and K is the union of finitely many intervals, then (f_N^*) converges to f^* weakly* in $L^\infty(\partial\mathbb{D})$. If $p = 1$ and $K \neq \partial\mathbb{D}$, then (f_N^*) converges to f^* weakly in $L^1(\partial\mathbb{D})$ and strongly in $L^1(\partial\mathbb{D} \setminus K)$.*

Proof. The sequence (f_N^*) is bounded in $L^\infty(\partial\mathbb{D})$. If $p = \infty$, then there is a weakly* convergent subsequence $(f_{N_l}^*)$. If $1 \le p < \infty$, then there is a subsequence $(f_{N_l}^*)$ which converges weakly in $L^p(\partial\mathbb{D})$: In the case $1 < p < \infty$ this is due to the fact that $(f_{N_l}^*)$ is especially bounded in $L^p(\partial\mathbb{D})$ and that the unit ball in reflexive spaces in weakly sequentially compact. In the case $p = 1$ we notice that, since (f_N^*) is especially bounded in $L^2(\partial\mathbb{D})$, we can extract a subsequence $(f_{N_l}^*)$ which converges weakly in $L^2(\partial\mathbb{D})$. But weak convergence in $L^2(\partial\mathbb{D})$ implies weak convergence $L^1(\partial\mathbb{D})$. Denote the limit in any case by $\widetilde{f}$.

Because the norm is sequentially lower semicontinuous with respect to weak and weak* convergence,

$$\|\widetilde{f} - \varphi\|_{L^p(K)} \le \liminf_{l\to\infty} \|f_{N_l}^* - \varphi\|_{L^p(K)} = \|f^* - \varphi\|_{L^p(K)}.$$

Equality on the right hand side follows from Theorem 5.1, for $p = \infty$ together with Theorem 4.15. The set of functions that is feasible for (H-OPT$_p$), $\{f \in H^\infty(\mathbb{D}) : |f| \le g \text{ on } \partial\mathbb{D}\}$, is weakly closed in $L^p(\partial\mathbb{D})$ for $1 \le p < \infty$, and (sequentially) weakly* closed in $L^\infty(\partial\mathbb{D})$. Therefore, the weak (or weak*) limit $\widetilde{f}$ is also feasible for (H-OPT$_p$), whence $\|\widetilde{f} - \varphi\|_{L^p(K)} = \|f^* - \varphi\|_{L^p(K)}$.

Uniqueness of the solution of (H-OPT$_p$) now implies $\widetilde{f} = f^*$. (In the case $p = 1$ we need $K \neq \partial\mathbb{D}$ for uniqueness.) But then it follows that the whole sequence (f_N^*) converges weakly (or weakly*) to f^*: If there were infinitely many f_N^* outside of an arbitrary (weak $L^p(\partial\mathbb{D})$- or weak* $L^\infty(\partial\mathbb{D})$-)neighborhood of f^*, we could use the preceding arguments to find a subsequence of these infinitely many f_N^* that converges to f^*, which is a contradiction.

If $1 < p < \infty$, then weak convergence, $f_N^* - \varphi \rightharpoonup f^* - \varphi$ in $L^p(K)$, and convergence of the norm, $\|f_N^* - \varphi\|_{L^p(K)} \to \|f^* - \varphi\|_{L^p(K)}$, imply that $f_N^* - \varphi \to f^* - \varphi$ strongly in $L^p(K)$ (see [3, Ü6.6]), and therefore $f_N^* \to f^*$ strongly in $L^p(K)$. Further, by Theorem 4.4, $|f^*| = g$ a.e. on $\partial\mathbb{D} \setminus K$. Then,

$$\|g\|_{L^p(\partial\mathbb{D}\setminus K)} = \|f^*\|_{L^p(\partial\mathbb{D}\setminus K)} \le \liminf_{l\to\infty} \|f_N^*\|_{L^p(\partial\mathbb{D}\setminus K)} \le \|g\|_{L^p(\partial\mathbb{D}\setminus K)},$$

from which it follows that $\|f_N^*\|_{L^p(\partial\mathbb{D}\setminus K)} \to \|g\|_{L^p(\partial\mathbb{D}\setminus K)}$. As before, weak convergence and convergence of the norm imply that $f_N^* \to f^*$ strongly in $L^p(\partial\mathbb{D} \setminus K)$. Together, $f_N^* \to f^*$ strongly in $L^p(\partial\mathbb{D})$ for $1 < p < \infty$.

Finally, because $|f^*| = g$ a.e. on $\partial\mathbb{D} \setminus K$ by Theorem 4.4 and because $|f_N^*| \le g$ for all N, weak convergence $f_N^* \rightharpoonup f^*$ in $L^1(\partial\mathbb{D} \setminus K)$ implies strong convergence $f_N^* \to f^*$ in $L^1(\partial\mathbb{D} \setminus K)$, see, e.g., [64, Theorem 1]. □

5.1.3 Fully discrete problem

Unfortunately, we are not aware of any "nice" method to check the constraint $|f| \le g$ on the complete circle. For a complete discretization we only check the constraint on some grid $\mathcal{X} \subset \partial\mathbb{D}$. Moreover, it may not be possible to compute the objective function exactly. We therefore replace $\|f - \varphi\|_{L^p(K)}$ by some quadrature approximation $T^p(f - \varphi)$ and obtain the fully discrete problem

$$\begin{array}{ll} \text{minimize} & T^p(f-\varphi) \\ \text{subject to} & |f(\mathrm{e}^{i\vartheta})| \le g(\mathrm{e}^{i\vartheta}), \quad \vartheta \in \mathcal{X}, \\ & f \in \mathcal{H}_N^\infty(\mathbb{D}). \end{array} \tag{FDP$_p$}$$

If we want to denote the above problem with, e.g., a specific grid $\mathcal{X}$, a specific approximation T^p, or a specific N, we write (FDP$_p(\mathcal{X})$), (FDP$_p(\mathcal{X}, T^p)$), (FDP$_p(\mathcal{X}, T^p, N)$) and so on. The set of feasible functions $\{f \in \mathcal{H}_N^\infty(\mathbb{D}) : |f(\mathrm{e}^{i\vartheta})| \le g(\mathrm{e}^{i\vartheta}),\ \vartheta \in \mathcal{X}\}$ is convex and closed in the finite dimensional space $\mathcal{H}_N^\infty(\mathbb{D})$. From Lemma 5.7 below it follows that if the grid $\mathcal{X}$ is fine enough, then the set of feasible functions is also bounded and therefore, because $\mathcal{H}_N^\infty(\mathbb{D})$ is finite dimensional, compact. Thus, (FDP$_p$) has a solution. If additionally the quadrature approximation is strictly convex, then the solution is unique.

We assume that we are given a sequence (T_n^p) of quadrature approximations that converges locally uniformly for functions $f \in \mathcal{H}_N^\infty(\mathbb{D})$, i.e.,

$$\sup_{f \in \mathcal{H}_N^\infty(\mathbb{D}),\ \|f\|_{L^\infty(\partial\mathbb{D})} \le 1} \left| T_n^p(f-\varphi) - \|f-\varphi\|_{L^p(K)} \right| \to 0 \qquad \text{as } n \to \infty. \tag{5.1}$$

Our aim is to show that, as the grid $\mathcal{X}$ on which we check the constraint becomes finer and as the approximation T^p becomes better, the minimum of the fully discrete problem (FDP$_p$) converges to the minimum of the semi-discrete problem (SDP$_p$):

Theorem 5.4. *Let $1 \le p \le \infty$ and fix $N \in \mathbb{N}$. Let $(\mathcal{X}_n)$ be a sequence of grids on $\partial\mathbb{D}$ with $h_{\max}(\mathcal{X}_n) \to 0$ as $n \to \infty$, and let (T_n^p) be a sequence of quadrature approximations to $\|\cdot\|_{L^p(K)}$ such that (5.1) holds. Then the minimum of the fully discrete problem (FDP$_p(\mathcal{X}_n, T_n^p, N)$) converges to the minimum of the semi-discrete problem (SDP$_p(N)$) as $n \to \infty$, i.e., if $f_{N,n}^*$ is a solution of (FDP$_p(\mathcal{X}_n, T_n^p, N)$), and f_N^* is a solution of (SDP$_p(N)$), then*

$$T_n^p(f_{N,n}^* - \varphi) \to \|f_N^* - \varphi\|_{L^p(K)} \qquad \text{as } n \to \infty.$$

Theorem 5.1 and Theorem 5.4 together imply

Theorem 5.5. *Let $(\mathcal{X}_n)$ be a sequence of grids on $\partial\mathbb{D}$ with $h_{\max}(\mathcal{X}_n) \to 0$ as $n \to \infty$, and let (T_n^p) be a sequence of quadrature approximations to $\|\cdot\|_{L^p(K)}$ such that (5.1) holds.*

(a) *Let* $1 \leq p < \infty$. *Then for each* $N \in \mathbb{N}$ *we can choose* $n(N)$ *such that the minimum of the fully discrete problem* $(\mathrm{FDP}_p(\mathcal{X}_{n(N)}, T^p_{n(N)}, N))$ *converges to the minimum of* (H-OPT_p) *as* $N \to \infty$.

(b) *Let* $p = \infty$. *Then for each* $N \in \mathbb{N}$ *we can choose* $n(N)$ *such that the minimum of the fully discrete problem* $(\mathrm{FDP}_p(\mathcal{X}_{n(N)}, T^p_{n(N)}, N))$ *converges to the infimum of* $(\mathcal{A}\text{-OPT}_p)$ *as* $N \to \infty$.

Before we can prove Theorem 5.4 we need a lemma which gives a bound on the derivative of functions that are feasible for $(\mathrm{FDP}_p(\mathcal{X}, N))$.

Lemma 5.6. *Fix* $N \in \mathbb{N}$. *There are* $h_0 > 0$ *and* $C > 0$ *such that for any grid* $\mathcal{X}$ *with* $h_{\max}(\mathcal{X}) \leq h_0$ *and any* f *which is feasible for* $(\mathrm{FDP}_p(\mathcal{X}, N))$ *the estimate*

$$\|f'\|_{L^\infty(\partial\mathbb{D})} \leq C\|g\|_{L^\infty(\partial\mathbb{D})}.$$

holds true. Here, $f'(\mathrm{e}^{i\vartheta}) = \frac{\mathrm{d}}{\mathrm{d}\vartheta} f(\mathrm{e}^{i\vartheta})$.

Proof. Let $\mathcal{X} = \{\mathrm{e}^{i\vartheta_1}, \ldots, \mathrm{e}^{i\vartheta_n}\} \subset \partial\mathbb{D}$ be some grid. For a coefficient vector $\beta = (\beta_{-N}, \beta_{-N+1}, \ldots, \beta_{N-1})^\top \in \mathbb{R}^{2N}$ we write the relations

$$f_\beta(\mathrm{e}^{i\vartheta_j}) = \sum_{k=-N}^{N-1} \beta_k \mathrm{e}^{ik\vartheta_j}, \qquad j = 1, \ldots, n,$$

in matrix form

$$B(\mathcal{X})\beta = f(\mathcal{X}). \tag{5.2}$$

Here, $f(\mathcal{X})$ is the vector $f(\mathcal{X}) = (f_\beta(\mathrm{e}^{i\vartheta_1}), \ldots, f_\beta(\mathrm{e}^{i\vartheta_n}))^\top \in \mathbb{C}^n$ and $B(\mathcal{X})$ is the matrix $B(\mathcal{X}) = (b_{jk}) \in \mathbb{C}^{n \times 2N}$, where $b_{jk} = \mathrm{e}^{ik\vartheta_j}$, $j = 1, \ldots, n$, $k = -N, \ldots, N-1$.

Now let $\vartheta_j = j\pi/N$ and take the grid $\mathcal{X}^N = \{\mathrm{e}^{i\vartheta_j} : j = 1, \ldots, 2N\}$ with $n = 2N$ points. Then $\frac{1}{\sqrt{2N}} B(\mathcal{X}^N)$ is unitary. Since the set of invertible matrices is open, and since matrix inversion is a continuous function on this set, there is $h_1 > 0$ such that if the grid $\widetilde{\mathcal{X}} = \{\mathrm{e}^{i\widetilde{\vartheta}_1}, \ldots, \mathrm{e}^{i\widetilde{\vartheta}_{2N}}\}$ satisfies

$$\max_{j=1,\ldots,2N} |\mathrm{e}^{i\vartheta_j} - \mathrm{e}^{i\widetilde{\vartheta}_j}| \leq h_1, \tag{5.3}$$

then $B(\widetilde{\mathcal{X}})$ is still invertible and

$$\|B(\widetilde{\mathcal{X}})^{-1}\|_{\infty\to 1} \leq 2\|B(\mathcal{X}^N)^{-1}\|_{\infty\to 1}. \tag{5.4}$$

Here, $\|\cdot\|_{\infty\to 1}$ is the operator norm $\|A\|_{\infty\to 1} = \sup_{\|x\|_\infty \le 1} \|Ax\|_1$.

To finish the proof, choose $h_0 > 0$ so small that any grid $\mathcal{X} \subset \partial\mathbb{D}$ with $h_{\max}(\mathcal{X}) \le h_0$ has a subgrid $\widetilde{\mathcal{X}} \subset \mathcal{X}$, consisting of $2N$ points, that satisfies (5.3). Let $\mathcal{X}$ be such a grid and $\widetilde{\mathcal{X}}$ a subgrid with (5.3). Let f be feasible for (FDP$_p(\mathcal{X}, N)$), i.e., $f \in \mathcal{H}^\infty_N(\mathbb{D})$ with $|f(\mathrm{e}^{i\vartheta})| \le g(\mathrm{e}^{i\vartheta})$ for all $\mathrm{e}^{i\vartheta} \in \mathcal{X}$. Then f has the form $f = f_\alpha$ for some $\alpha \in \mathbb{R}^N$. By (5.2) we have

$$B(\widetilde{\mathcal{X}}) \begin{pmatrix} 0_{\mathbb{R}^N} \\ \alpha \end{pmatrix} = f(\widetilde{\mathcal{X}}),$$

whence

$$\begin{aligned} \|\alpha\|_1 &\le \|B(\widetilde{\mathcal{X}})^{-1}\|_{\infty\to 1} \|f(\widetilde{\mathcal{X}})\|_\infty = \|B(\widetilde{\mathcal{X}})^{-1}\|_{\infty\to 1} \max_{\mathrm{e}^{i\widetilde{\vartheta}} \in \widetilde{\mathcal{X}}} |f(\mathrm{e}^{i\widetilde{\vartheta}})| \\ &\overset{(5.4)}{\le} 2\|B(\mathcal{X}^N)^{-1}\|_{\infty\to 1} \|g\|_{L^\infty(\partial\mathbb{D})}. \end{aligned}$$

Therefore,

$$\|f'\|_{L^\infty(\partial\mathbb{D})} = \sup_{\mathrm{e}^{i\vartheta} \in \partial\mathbb{D}} \left| \sum_{k=1}^{N-1} ik\alpha_k \mathrm{e}^{ik\vartheta} \right| \le N\|\alpha\|_1 \le 2N\|B(\mathcal{X}^N)^{-1}\|_{\infty\to 1} \|g\|_{L^\infty(\partial\mathbb{D})}.$$

Thus, the lemma holds with $C = 2N\|B(\mathcal{X}^N)^{-1}\|_{\infty\to 1}$. □

Next, we show that if the grid $\mathcal{X}_n$ is fine enough, then functions that are feasible for the fully discrete problem (FDP$_p(\mathcal{X}_n, N)$) are almost feasible for the semi-discrete problem (SDP$_p(N)$).

Lemma 5.7. *Fix $N \in \mathbb{N}$. Then for any $\epsilon > 0$ there is $h_1 > 0$ such that if $\mathcal{X}$ is a grid with $h_{\max}(\mathcal{X}) \le h_1$ and f is feasible for* (FDP$_p(\mathcal{X}, N)$), *then*

$$|f| \le g + \epsilon.$$

Proof. Since g is uniformly continuous, there is $h_2 > 0$ such that for all $|\mu| \le h_2$ and all $\mathrm{e}^{i\vartheta} \in \partial\mathbb{D}$ we have

$$|g(\mathrm{e}^{i\vartheta}) - g(\mathrm{e}^{i(\vartheta+\mu)})| \le \epsilon/2. \tag{5.5}$$

Suppose that $\mathcal{X}$ is a grid with

$$h_{\max}(\mathcal{X}) \le \min\{h_0, h_2, \epsilon(C\|g\|_{L^\infty(\partial\mathbb{D})})^{-1}\}, \tag{5.6}$$

where h_0 and C are the constants from Lemma 5.6. Let f be feasible for $(\mathrm{FDP}_p(\mathcal{X}, N))$ and let $\mathrm{e}^{i\vartheta} \in \partial\mathbb{D}$. Choose $\mathrm{e}^{it} \in \mathcal{X}$ such that $\operatorname{dist}\left(\mathrm{e}^{i\vartheta}, \mathrm{e}^{it}\right) \le h_{\max}(\mathcal{X})/2$. Then

$$\begin{aligned}
|f(\mathrm{e}^{i\vartheta})| &\le |f(\mathrm{e}^{it})| + \frac{h_{\max}(\mathcal{X})}{2}\|f'\|_{L^\infty(\partial\mathbb{D})} \\
&\le g(\mathrm{e}^{it}) + \frac{h_{\max}(\mathcal{X})}{2} C\|g\|_{L^\infty(\partial\mathbb{D})} && \text{by Lemma 5.6} \\
&\le g(\mathrm{e}^{i\vartheta}) + \frac{\epsilon}{2} + \frac{\epsilon}{2} && \text{by (5.5) and (5.6)} \\
&= g(\mathrm{e}^{i\vartheta}) + \epsilon.
\end{aligned}$$

Thus, the lemma holds true with $h_1 = \min\{h_0, h_2, \epsilon(C\|g\|_{L^\infty(\partial\mathbb{D})})^{-1}\}$. □

We are now ready to prove Theorem 5.4.

Proof of Theorem 5.4. Fix an arbitrary $\epsilon > 0$. Let $\epsilon_1 > 0$ be so small that for

$$\eta = \left(1 + \frac{\epsilon_1}{\min_{\mathrm{e}^{i\vartheta}\in\partial\mathbb{D}} g(\mathrm{e}^{i\vartheta})}\right)^{-1} \tag{5.7}$$

it holds true that $(1-\eta)\|\varphi\|_{L^p(K)} \le \epsilon$. (This is possible since $\eta \to 1$ as $\epsilon_1 \to 0$.) By Lemma 5.7 there is $n_0 \in \mathbb{N}$ such that functions that are feasible for $(\mathrm{FDP}_p(\mathcal{X}_n, N))$, $n \ge n_0$, satisfy

$$|f| \le g + \epsilon_1. \tag{5.8}$$

Because of (5.1) we can possibly increase n_0 such that for all $n \ge n_0$ and all $f \in \mathcal{H}_N^\infty(\mathbb{D})$ with $\|f\|_{L^\infty(\partial\mathbb{D})} \le \|g\|_{L^\infty(\partial\mathbb{D})} + \epsilon_1$

$$\left|T_n^p(f-\varphi) - \|f-\varphi\|_{L^p(K)}\right| \le \epsilon. \tag{5.9}$$

So (5.9) especially holds true for all f that are feasible for $(\mathrm{FDP}_p(\mathcal{X}_n, N))$, $n \ge n_0$.

Now fix $n \ge n_0$. Let $f_{N,n}^*$ be a solution of the fully discrete problem $(\mathrm{FDP}_p(\mathcal{X}_n, T_n^p, N))$, and let f_N^* be a solution of the semi-discrete problem $(\mathrm{SDP}_p(N))$. Then

$$T_n^p(f_{N,n}^* - \varphi) \le T_n^p(f_N^* - \varphi) \le \|f_N^* - \varphi\|_{L^p(K)} + \epsilon. \tag{5.10}$$

The first inequality holds true because $f_{N,n}^*$ is a solution of $(\mathrm{FDP}_p(\mathcal{X}_n, T_n^p, N))$ and f_N^* is feasible for $(\mathrm{FDP}_p(\mathcal{X}_n, T_n^p, N))$, and the second inequality is due to (5.9).

On the other hand, take η from (5.7) and set $f_{N,n} = \eta f^*_{N,n}$. Because $f^*_{N,n}$ is feasible for $(\mathrm{FDP}_p(\mathcal{X}_n, N))$, we have $|f^*_{N,n}| \le g + \epsilon_1$ by (5.8), and therefore

$$|f^*_{N,n}| \le \left(1 + \frac{\epsilon_1}{\min_{e^{i\vartheta}\in\partial\mathbb{D}} g(e^{i\vartheta})}\right) g.$$

So $f_{N,n} = \eta f^*_{N,n} \le g$, that is, $f_{N,n}$ is feasible for the semi-discrete problem. Therefore,

$$\|f^*_N - \varphi\|_{L^p(K)} \le \|f_{N,n} - \varphi\|_{L^p(K)}.$$

Further, since $\eta \le 1$ and $(1-\eta)\|\varphi\|_{L^p(K)} \le \epsilon$,

$$\begin{aligned}\|f_{N,n} - \varphi\|_{L^p(K)} &= \|\eta f^*_{N,n} - \varphi\|_{L^p(K)} \le \eta\|f^*_{N,n} - \varphi\|_{L^p(K)} + (1-\eta)\|\varphi\|_{L^p(K)} \\ &\le \|f^*_{N,n} - \varphi\|_{L^p(K)} + \epsilon.\end{aligned}$$

Because (5.9) holds true for $f^*_{N,n}$, we end up with

$$\|f^*_N - \varphi\|_{L^p(K)} \le T^p_n(f^*_{N,n} - \varphi) + 2\epsilon \tag{5.11}$$

for $n \ge n_0$.

(5.10) and (5.11) imply that $T^p_n(f^*_{N,n} - \varphi) \to \|f^*_N - \varphi\|_{L^p(K)}$ as $n \to \infty$. □

Remark 5.8. *Theorem 5.4 still holds true for more general g. We used the continuity of g only to prove Lemma 5.7, which we used in the proof of Theorem 5.4. One can show that Lemma 5.7 is still true if, for example, g is continuous up to finitely many jump discontinuities.*

Corollary 5.9. *Let $1 < p < \infty$ and fix $N \in \mathbb{N}$. Let $(\mathcal{X}_n)$ be a sequence of grids on $\partial\mathbb{D}$ with $h_{\max}(\mathcal{X}_n) \to 0$ as $n \to \infty$, and let (T^p_n) be a sequence of quadrature approximations to $\|\cdot\|_{L^p(K)}$ such that (5.1) holds. Let $f^*_{N,n}$ be a solution of $(\mathrm{FDP}_p(\mathcal{X}_n, T^p_n, N))$, and let f^*_N be the solution of $(\mathrm{SDP}_p(N))$. Then $(f^*_{N,n})$ converges to f^*_N strongly in $L^p(\partial\mathbb{D})$ as $n \to \infty$.*

We omit the proof, because it is similar to the proof of Corollary 5.3. We instead prove

Theorem 5.10. *Let $(\mathcal{X}_n)$ be a sequence of grids on $\partial\mathbb{D}$ with $h_{\max}(\mathcal{X}_n) \to 0$ as $n \to \infty$, and let (T^p_n) be a sequence of quadrature approximations to $\|\cdot\|_{L^p(K)}$ such that (5.1) holds. Let f^* be a solution of $(\mathrm{H\text{-}OPT}_p)$ and let $f^*_{N,n}$ be a solution of $(\mathrm{FDP}_p(\mathcal{X}_n, T^p_n, N))$.*

(a) *If* $1 < p < \infty$, *then for each* $N \in \mathbb{N}$ *we can choose* $n(N)$ *such that the minimum of* ($\mathrm{FDP}_p(\mathcal{X}_{n(N)}, T^p_{n(N)}, N)$) *converges to the minimum of* (H-OPT_p) *and such that* $f^*_{N,n(N)} \to f^*$ *in* $L^p(\partial\mathbb{D})$ *as* $N \to \infty$.

(b) *If* $p = \infty$ *and* K *is the union of finitely many closed intervals, then for each* $N \in \mathbb{N}$ *we can choose* $n(N)$ *such that the minimum of the discrete problem* ($\mathrm{FDP}_p(\mathcal{X}_{n(N)}, T^p_{n(N)}, N)$) *converges to the minimum of* (H-OPT_p) *and such that* $f^*_{N,n(N)} \overset{*}{\rightharpoonup} f^*$ *in* $L^\infty(\partial\mathbb{D})$ *as* $N \to \infty$.

(c) *If* $p = 1$, *then for each* $N \in \mathbb{N}$ *we can choose* $n(N)$ *such that the minimum of the discrete problem* ($\mathrm{FDP}_p(\mathcal{X}_{n(N)}, T^p_{n(N)}, N)$) *converges to the minimum of* (H-OPT_p), *and, if* $K \neq \partial\mathbb{D}$, *such that* $f^*_{N,n(N)} \rightharpoonup f^*$ *in* $L^1(\partial\mathbb{D})$ *and* $f^*_{N,n(N)} \to f^*$ *in* $L^1(\partial\mathbb{D} \setminus K)$ *as* $N \to \infty$.

Proof. By Theorem 5.5, in the case $p = \infty$ together with Theorem 4.15, we can choose $n(N)$ such that the minimum of ($\mathrm{FDP}_p(\mathcal{X}_n, T^p_n, N)$) converges to the minimum of (H-OPT_p). As we saw in the proof of Theorem 5.4, we can additionally achieve that

$$|T^p_n(f^*_{N,n(N)} - \varphi) - \|f^*_{N,n(N)} - \varphi\|_{L^p(K)}| \leq \epsilon_N \tag{5.12}$$

and $|f^*_{N,n(N)}| \leq g + \epsilon_N$ with $\epsilon_N \to 0$ as $N \to \infty$. Especially, $(f^*_{N,n(N)})$ is bounded in $L^\infty(\partial\mathbb{D})$. As in the proof of Corollary 5.3 we can extract a subsequence which converges weakly in the case $1 \leq p < \infty$ and weakly* in the case $p = \infty$. Denote the limit by $\widetilde{f}$. As before, lower semicontinuity of the norm implies $\|\widetilde{f} - \varphi\|_{L^p(K)} \leq \|f^* - \varphi\|_{L^p(K)}$.

Now for any $\epsilon > 0$ all but possibly finitely many $f^*_{N,n(N)}$ lie in the set $\{f \in H^\infty(\mathbb{D}) : |f| \leq g + \epsilon \text{ on } \partial\mathbb{D}\}$. This set is weakly closed in $L^p(\mathbb{D})$, $1 \leq p < \infty$, and (sequentially) weakly* closed in $L^\infty(\mathbb{D})$. Therefore, $|\widetilde{f}| \leq g + \epsilon$ for any $\epsilon > 0$, i.e., $|\widetilde{f}| \leq g$. But this means that $\widetilde{f}$ is feasible for (H-OPT_p). If follows that $\widetilde{f}$ is a solution of (H-OPT_p). Unique solvability then implies $\widetilde{f} = f^*$. (In the case $p = 1$ we need $K \neq \partial\mathbb{D}$.) As before, by uniqueness of the limit the whole sequence $(f^*_{N,n(N)})$ converges weakly (or weakly* if $p = \infty$) to f^* in $L^p(\partial\mathbb{D})$ as $N \to \infty$.

Because of (5.12) and because $T^p_n(f^*_{N,n(N)} - \varphi)$ converges to $\|f^* - \varphi\|_{L^p(K)}$, $\|f^*_{N,n(N)} - \varphi\|_{L^p(K)} \to \|f^* - \varphi\|_{L^p(K)}$. Also, it follows as in the proof of Corollary 5.3 that $\|f^*_{N,n(N)}\|_{L^p(\partial\mathbb{D}\setminus K)} \to \|f^*\|_{L^p(\partial\mathbb{D}\setminus K)}$. As before, we obtain that $f^*_{N,n(N)} \to f^*$ strongly in $L^p(\partial\mathbb{D})$ for $1 < p < \infty$. In the case $p = 1$, weak convergence $f^*_{N,n(N)} \rightharpoonup f^*$ in $L^1(\partial\mathbb{D}\setminus K)$, together with the properties $|f^*_{N,n(N)}| \le g + \epsilon_N$ with $\epsilon_N \to 0$ and $|f^*| = g$ a.e. on $\partial\mathbb{D}\setminus K$, implies strong convergence $f^*_{N,n(N)} \to f^*$ in $L^1(\partial\mathbb{D}\setminus K)$, see, e.g., [64, Lemma 2]. □

Example 5.11 (Rectangle rule, case $1 \le p < \infty$). *Assume that K is the union of finitely many closed intervals of positive measure and that φ is smooth. Let $(\mathcal{X}_n)$ be a sequence of grids with $h_{\max}(\mathcal{X}_n) \to 0$ as $n \to \infty$. Then (5.1) is fulfilled for the rectangle rule*

$$T_n^p(f-\varphi) = \left(\sum_{e^{i\vartheta}\in K\cap\mathcal{X}_n} \left| f(e^{i\vartheta}) - \varphi(e^{i\vartheta}) \right|^p h_\vartheta \right)^{1/p},$$

where

$$h_\vartheta = \frac{\min\{\mu>0 : e^{i(\vartheta-\mu)} \in \mathcal{X}\} + \min\{\mu>0 : e^{i(\vartheta+\mu)} \in \mathcal{X}\}}{2}.$$

The reason why (5.1) holds for the rectangle rule is Lemma 5.6 and the fact that the error of the rectangle rule can be estimated by the derivative of the integrand.

Example 5.12 (Exact quadrature, case $p=2$). *We will see in the next section that for $p=2$, $\varphi \in \mathcal{L}^\infty_{N_\varphi}(\partial\mathbb{D})$ for some $N_\varphi \in \mathbb{N}$ and $f \in \mathcal{H}^\infty_N(\mathbb{D})$ it is in principle possible to compute $\|f-\varphi\|_{L^p(K)}$ exactly. Assume that (φ_n) is a sequence with $\varphi_n \in \mathcal{L}^\infty_n(\partial\mathbb{D})$ and $\|\varphi_n - \varphi\|_{L^2(K)} \to \infty$ as $n\to\infty$ and suppose that*

$$T_n^2(f-\varphi) = \|f-\varphi_n\|_{L^2(K)}.$$

Then

$$\begin{aligned}\left|T_n^2(f-\varphi) - \|f-\varphi\|_{L^2(K)}\right| &= \left|\|f-\varphi_n\|_{L^2(K)} - \|f-\varphi\|_{L^2(K)}\right| \\ &\le \|\varphi-\varphi_n\|_{L^2(K)},\end{aligned}$$

so (T_n^2) satisfies (5.1).

Example 5.13 (Case $p=\infty$). *Assume that K is the union of finitely many closed intervals of positive measure and that φ is smooth. Let $(\mathcal{X}_n)$ be a sequence of grids with $h_{\max}(\mathcal{X}_n) \to 0$ as $n\to\infty$. Then (5.1) is fulfilled for*

$$T_n^p(f-\varphi) = \max_{e^{i\vartheta}\in K\cap\mathcal{X}_n} |f(e^{i\vartheta}) - \varphi(e^{i\vartheta})|.$$

The reason why (5.1) holds is again Lemma 5.6.

5.2 Discretization: Examples

In this section we consider different discretizations of (H-OPT$_p$) corresponding to different approximations of the objective functions.

5.2.1 $1 \le p < \infty$, rectangle rule

Let $\mathcal{X} = \{\mathrm{e}^{i\vartheta_1}, \ldots, \mathrm{e}^{i\vartheta_d}\}$ and let $N \in \mathbb{N}$. With the rectangle rule to approximate $\|f - \varphi\|_{L^2(K)}$ (see Example 5.11) we get the problem

$$\begin{array}{ll} \text{minimize} & \left(\sum_{j \in K \cap \mathcal{X}} \left| f_\alpha(\mathrm{e}^{i\vartheta_j}) - \varphi_j \right|^p h_{\vartheta_j} \right)^{1/p} \\ \text{subject to} & \left| f_\alpha(\mathrm{e}^{i\vartheta_j}) \right| \le g_j, \qquad j \in \mathcal{X}, \end{array}$$

in the optimization variable $\alpha \in \mathbb{R}^N$. For the definition of f_α see Section 5.1.1. The sloppy notation $j \in \mathcal{X}$ means $\mathrm{e}^{i\vartheta_j} \in \mathcal{X}$, similarly for $j \in K \cap \mathcal{X}$. Also, we write $\varphi_j = \varphi(\mathrm{e}^{i\vartheta_j})$ and $g_j = g(\mathrm{e}^{i\vartheta_j})$, $j = 1, \ldots, d$.

Let us assume that $\mathcal{X}$ is symmetric, i.e., $\mathcal{X} = \overline{\mathcal{X}}$, and write

$$\mathcal{X}^+ = \mathcal{X} \cap \{\mathrm{e}^{i\vartheta} : \vartheta \in [0, \pi]\}.$$

Then due to the symmetry of f_α, g, φ and K it actually suffices to consider

$$\begin{array}{lll} \text{minimize} & F_p(\alpha) = \sum_{j \in K \cap \mathcal{X}} \left| f_\alpha(\mathrm{e}^{i\vartheta_j}) - \varphi_j \right|^p h_{\vartheta_j} & \\ \text{subject to} & G_j(\alpha) = \left| f_\alpha(\mathrm{e}^{i\vartheta_j}) \right|^2 - g_j^2 \le 0, \qquad j \in \mathcal{X}^+. & \end{array} \tag{D-OPT$_p$}$$

For practical purposes it is useful to drop the $1/p$-th power from the objective function and to write the constraints as in (D-OPT$_p$): For $p = 2$, (D-OPT$_p$) is a quadratically constrained quadratic program (QCQP). We will write down the QCQP formulation of (D-OPT$_2$) more explicitly in Section 5.3.

5.2.2 $p = 2$, exact quadrature

We come back to Example 5.12. Suppose that φ is nice enough and can be written (or well approximated) in the form

$$\varphi(\mathrm{e}^{i\vartheta}) = \sum_{k=-N_\varphi}^{N_\varphi - 1} \beta_k \mathrm{e}^{ik\vartheta}$$

with some $N_\varphi \in \mathbb{N}$ and $\beta = (\beta_k)_{k=-N_\varphi}^{N_\varphi - 1} \in \mathbb{R}^{2N_\varphi}$. (The β_k must be real due to the real symmetry of φ.) Then it is possible to compute $\|f_\alpha - \varphi\|_{L^2(K)}$ exactly.

Without loss of generality we may assume that $N_\varphi \geq N$ with possibly some of the β_k equal to zero. To simplify notation in the following, we write $f_\alpha(\mathrm{e}^{i\vartheta}) = \sum_{k=-N_\varphi}^{N_\varphi-1} \alpha_k \mathrm{e}^{ik\vartheta}$ with $\alpha_k = 0$ for $k \notin \{0, 1, \ldots, N-1\}$. We have

$$\begin{aligned}
\|f_\alpha - \varphi\|_{L^2(K)}^2 &= \int_K |f_\alpha(\mathrm{e}^{i\vartheta}) - \varphi(\mathrm{e}^{i\vartheta})|^2 \,\mathrm{d}\vartheta = \int_K \left| \sum_{k=-N_\varphi}^{N_\varphi-1} (\alpha_k - \beta_k)\mathrm{e}^{ik\vartheta} \right|^2 \mathrm{d}\vartheta \\
&= \int_K \sum_{k,l=-N_\varphi}^{N_\varphi-1} (\alpha_k - \beta_k)(\alpha_l - \beta_l)\mathrm{e}^{i(k-l)\vartheta} \,\mathrm{d}\vartheta \\
&= \sum_{k,l=-N_\varphi}^{N_\varphi-1} (\alpha_k - \beta_k)(\alpha_l - \beta_l) \int_K \mathrm{e}^{i(k-l)\vartheta} \,\mathrm{d}\vartheta.
\end{aligned}$$

With

$$m_j = \int_K \mathrm{e}^{ij\vartheta} \,\mathrm{d}\vartheta$$

we get the discrete problem

$$\begin{aligned}
&\text{minimize} && \widetilde{F}_2(\alpha) = \sum_{k,l=-N_\varphi}^{N_\varphi-1} m_{k-l}(\alpha_k - \beta_k)(\alpha_l - \beta_l) \\
&\text{subject to} && G_j(\alpha) = \left|f_\alpha(\mathrm{e}^{i\vartheta_j})\right|^2 - g_j^2 \leq 0, \qquad j \in \mathcal{X}^+.
\end{aligned} \tag{D-OPT$_{\widetilde{2}}$}$$

5.2.3 $p = \infty$

In the case $p = \infty$ we use the approximation from Example 5.13 and consider

$$\begin{aligned}
&\text{minimize} && F_\infty(\alpha) = \max_{j \in K \cap \mathcal{X}^+} \left|f_\alpha(\mathrm{e}^{i\vartheta_j}) - \varphi_j\right|^2 \\
&\text{subject to} && G_j(\alpha) = \left|f_\alpha(\mathrm{e}^{i\vartheta_j})\right|^2 - g_j^2 \leq 0, \qquad j \in \mathcal{X}^+.
\end{aligned} \tag{D-OPT$_\infty$}$$

Notice that due to symmetry we only take the maximum over $j \in K \cap \mathcal{X}^+$. The reason for the square is that this allows us to write (D-OPT$_\infty$) as a QCQP in the following section.

5.3 QCQP formulation of the discrete problems

In this (technical) section we write the problems (D-OPT$_2$), (D-OPT$_{\widetilde{2}}$) and (D-OPT$_\infty$) more explicitly as QCQPs. We will use this in the following sections when we recast these problems as second-order cone programs (SOCPs), for which there are efficient solvers available.

If $p \notin \{2, \infty\}$, but p is *rational*, then it is still possible to recast (D-OPT$_p$) as an SOCP. However, we do not treat this case, because the reformulation is not especially hard, but more cumbersome, and we want to spare the reader the technical details. For a strategy to obtain the SOCP formulation in this case, see, e.g., [2].

5.3.1 $p = 2$, rectangle rule

In the case $p = 2$ with the rectangle rule we have for the objective function

$$
\begin{aligned}
F_2(\alpha) &= \sum_{j \in K \cap \mathcal{X}} \left| f_\alpha(\mathrm{e}^{i\vartheta_j}) - \varphi_j \right|^2 h_{\vartheta_j} = \sum_{j \in K \cap \mathcal{X}} \left| \sum_{k=0}^{N-1} \alpha_k \mathrm{e}^{ik\vartheta_j} - \varphi_j \right|^2 h_{\vartheta_j} \\
&= \sum_{j \in K \cap \mathcal{X}} \left(\left| \sum_{k=0}^{N-1} \alpha_k \mathrm{e}^{ik\vartheta_j} \right|^2 - 2\,\mathrm{Re}\left(\sum_{k=0}^{N-1} \alpha_k \mathrm{e}^{-ik\vartheta_j} \varphi_j \right) + |\varphi_j|^2 \right) h_{\vartheta_j}.
\end{aligned}
$$

For the first summand we have

$$
\begin{aligned}
\sum_{j \in K \cap \mathcal{X}} \left| \sum_{k=0}^{N-1} \alpha_k \mathrm{e}^{ik\vartheta_j} \right|^2 h_{\vartheta_j} &= \sum_{j \in K \cap \mathcal{X}} \sum_{k=0}^{N-1} \sum_{l=0}^{N-1} \alpha_k \alpha_l \mathrm{e}^{i(k-l)\vartheta_j} h_{\vartheta_j} \\
&= \sum_{j \in K \cap \mathcal{X}} \alpha^\top \mathbf{e}_j \overline{\mathbf{e}_j}^\top \alpha h_{\vartheta_j} \\
&= \alpha^\top \left(\sum_{j \in K \cap \mathcal{X}} \mathbf{e}_j \overline{\mathbf{e}_j}^\top h_{\vartheta_j} \right) \alpha,
\end{aligned}
$$

where $\mathbf{e}_j = (\mathrm{e}^{ik\vartheta_j})_{k=0,\dots,N-1} \in \mathbb{C}^N$. Since the expression is real,

$$
\sum_{j \in K \cap \mathcal{X}} \left| \sum_{k=0}^{N-1} \alpha_k \mathrm{e}^{ik\vartheta_j} \right|^2 h_{\vartheta_j} = \alpha^\top \left(\mathrm{Re} \sum_{j \in K \cap \mathcal{X}} \mathbf{e}_j \overline{\mathbf{e}_j}^\top h_{\vartheta_j} \right) \alpha.
$$

For the second summand we have

$$
\begin{aligned}
-\sum_{j \in K \cap \mathcal{X}} 2\,\mathrm{Re}\left(\sum_{k=0}^{N-1} \alpha_k \mathrm{e}^{-ik\vartheta_j} \varphi_j \right) h_{\vartheta_j} &= -2 \sum_{k=0}^{N-1} \alpha_k \,\mathrm{Re}\left(\sum_{j \in K \cap \mathcal{X}} \mathrm{e}^{-ik\vartheta_j} \varphi_j h_{\vartheta_j} \right) \\
&= 2q^\top \alpha,
\end{aligned}
$$

where

$$q = (q_k)_{k=0,\dots,N-1}, \qquad q_k = -\operatorname{Re}\left(\sum_{j\in K\cap\mathcal{X}} \mathrm{e}^{-ik\vartheta_j}\varphi_j h_{\vartheta_j}\right).$$

It follows that

$$F_2(\alpha) = \alpha^\top \left(\operatorname{Re}\sum_{j\in K\cap\mathcal{X}} \mathbf{e}_j\overline{\mathbf{e}_j}^\top h_{\vartheta_j}\right)\alpha + 2q^\top\alpha + \left(\sum_{j\in K\cap\mathcal{X}} |\varphi_j|^2 h_{\vartheta_j}\right).$$

5.3.2 $p = 2$, exact quadrature

For the case $p = 2$ with exact quadrature we have (see last section)

$$\begin{aligned}\widetilde{F}_2(\alpha) &= \sum_{k,l=-N_\varphi}^{N_\varphi-1} m_{k-l}(\alpha_k-\beta_k)(\alpha_l-\beta_l)\\ &= \sum_{k,l=0}^{N-1} m_{k-l}\alpha_k\alpha_l - 2\sum_{k=0}^{N-1}\left(\sum_{l=-N_\varphi}^{N_\varphi-1} m_{k-l}\beta_l\right)\alpha_k + \sum_{k,l=-N_\varphi}^{N_\varphi-1} m_{k-l}\beta_k\beta_l,\end{aligned}$$

where $m_j = \int_K \mathrm{e}^{ij\vartheta}\,\mathrm{d}\vartheta$. Since $K = \overline{K}$ we have

$$m_j = \operatorname{Re}\left(\int_K \mathrm{e}^{ij\vartheta}\,\mathrm{d}\vartheta\right) = \int_K \cos(j\vartheta)\,\mathrm{d}\vartheta$$

and $m_j = m_{-j}$. We let

$$\widetilde{M} = (m_{k-l})_{k,l=0}^{N-1} \in \mathbb{R}^{N\times N}.$$

$\widetilde{M}$ is a symmetric Toeplitz matrix, that is, it is constant along lines parallel to the diagonal. Moreover, let

$$\widetilde{q} = (\widetilde{q}_k)_{k=0,\dots,N-1}, \qquad \widetilde{q}_k = -\sum_{l=-N_\varphi}^{N_\varphi-1} m_{k-l}\beta_l.$$

Then

$$\widetilde{F}_2(\alpha) = \alpha^\top\widetilde{M}\alpha + 2\widetilde{q}^\top\alpha + \sum_{k,l=-N_\varphi}^{N_\varphi-1} m_{k-l}\beta_k\beta_l.$$

5.3.3 $p = \infty$

In the case $p = \infty$ we have

$$F_\infty(\alpha) = \max_{j \in K \cap \mathcal{X}} \underbrace{\left| f_\alpha(\mathrm{e}^{i\vartheta_j}) - \varphi_j \right|^2}_{=: F^j(\alpha)}.$$

It turns out that

$$F^j(\alpha) = \alpha^\top \left(\mathrm{Re}\, \mathbf{e}_j \overline{\mathbf{e}_j}^\top \right) \alpha + 2(q^{(j)})^\top \alpha + |\varphi_j|^2,$$

where

$$q^{(j)} = (q_k^{(j)})_{k=0,\dots,N-1}, \qquad q_k^{(j)} = -\,\mathrm{Re}\left(\mathrm{e}^{-ik\vartheta_j} \varphi_j \right).$$

5.3.4 Constraints

For the constraints we have

$$\begin{aligned} G_j(\alpha) &= \left| f_\alpha(\mathrm{e}^{i\vartheta_j}) \right|^2 - g_j^2 = \left| \sum_{k=0}^{N-1} \alpha_k \mathrm{e}^{ik\vartheta_j} \right|^2 - g_j^2 \\ &= \left(\sum_{k=0}^{N-1} \alpha_k \mathrm{e}^{ik\vartheta_j} \right) \left(\sum_{l=0}^{N-1} \alpha_l \mathrm{e}^{-ik\vartheta_j} \right) - g_j^2 = \alpha^\top \mathbf{e}_j \overline{\mathbf{e}_j}^\top \alpha - g_j^2 \\ &= \alpha^\top \left(\mathrm{Re}\, \mathbf{e}_j \overline{\mathbf{e}_j}^\top \right) \alpha - g_j^2. \end{aligned}$$

Now $\mathbf{e}_j = \gamma_j + i\sigma_j$, where

$$\gamma_j = (\cos(k\vartheta_j))_{k=0,\dots,N-1} \quad \text{and} \quad \sigma_j = (\sin(k\vartheta_j))_{k=0,\dots,N-1}.$$

Hence,

$$\mathrm{Re}\, \mathbf{e}_j \overline{\mathbf{e}_j}^\top = \mathrm{Re}\left((\gamma_j + i\sigma_j)(\gamma_j - i\sigma_j)^\top \right) = \gamma_j \gamma_j^\top + \sigma_j \sigma_j^\top.$$

5.3.5 Summary

To summarize, the QCQP formulation for $p = 2$ with the rectangle rule is

$$\begin{aligned} &\text{minimize} \quad \alpha^\top \left(\sum_{j \in K \cap \mathcal{X}} (\gamma_j \gamma_j^\top + \sigma_j \sigma_j^\top) h_{\vartheta_j} \right) \alpha + 2q^\top \alpha + \left(\sum_{j \in K \cap \mathcal{X}} |\varphi_j|^2 h_{\vartheta_j} \right) \\ &\text{subject to} \quad \alpha^\top \left(\gamma_j \gamma_j^\top + \sigma_j \sigma_j^\top \right) \alpha - g_j^2 \le 0, \quad j \in \mathcal{X}^+ \end{aligned} \tag{QCQP$_2$}$$

in the optimization variable $\alpha \in \mathbb{R}^N$. For the case $p = 2$ with exact quadrature we have

$$\begin{aligned} &\text{minimize} \quad && \alpha^\top \widetilde{M} \alpha + 2\widetilde{q}^\top \alpha + \left(\sum_{k,l=-N_\varphi}^{N_\varphi - 1} m_{k-l} \beta_k \beta_l \right) \\ &\text{subject to} \quad && \alpha^\top \left(\gamma_j \gamma_j^\top + \sigma_j \sigma_j^\top \right) \alpha - g_j^2 \leq 0, \quad j \in \mathcal{X}^+ \end{aligned} \qquad (\mathrm{QCQP}_{\tilde{2}})$$

in the optimization variable $\alpha \in \mathbb{R}^N$. For $p = \infty$, the QCQP formulation is

$$\begin{aligned} &\text{minimize} \quad && t \\ &\text{subject to} \quad && \alpha^\top \left(\gamma_j \gamma_j^\top + \sigma_j \sigma_j^\top \right) \alpha + 2(q^{(j)})^\top \alpha + |\varphi_j|^2 - t \leq 0, \quad j \in K \cap \mathcal{X}^+, \\ & && \alpha^\top \left(\gamma_j \gamma_j^\top + \sigma_j \sigma_j^\top \right) \alpha - g_j^2 \leq 0, \quad j \in \mathcal{X}^+ \end{aligned} \qquad (\mathrm{QCQP}_\infty)$$

in the optimization variables $t \in \mathbb{R}$ and $\alpha \in \mathbb{R}^N$.

5.4 Second-order cone programs (SOCPs)

In principle, one could try to solve the discrete problems (D-OPT_p) (or (QCQP_p)) with general-purpose optimization methods like SQP [22]. However, this is only advisable for small N and $|\mathcal{X}^+|$. SQP becomes rather impractical for moderately large N and $|\mathcal{X}^+|$ due to the large number of non-sparse constraints in (QCQP_p). An SQP-based solver from the MATLAB optimization toolbox that we tried did not yield any usable results in reasonable time even for $|\mathcal{X}^+| = N = 256$.

During the 1980s and 1990s a class of methods that is much more efficient for certain convex optimization problems has been developed, *interior-point methods* [45]. Just like for finite element methods, their general theory constitutes a framework, and in practice a large number of right implementation choices has to be made in order to make them efficient [55, 63]. Although one can use interior-point methods to solve QCQPs directly, as it is for example done in [34], the more common approach is to recast QCQPs as *second-order cone programs (SOCPs)*.

The *standard second-order cone* (or *quadratic* or *Lorentz cone*) of dimension $n \in \mathbb{N}$ is the set

$$\mathcal{Q}_n = \left\{ (\xi_0; \xi) \in \mathbb{R} \times \mathbb{R}^{n-1} : \|\xi\|_2 \leq \xi_0 \right\}.$$

The semicolon is used to denote concatenation of vectors or matrices in a column, i.e., $(\xi_0; \xi) = \begin{pmatrix} \xi_0 \\ \xi \end{pmatrix}$. For $n = 1$ the definition has to be read as

$$\mathcal{Q}_1 = \{\xi_0 \in \mathbb{R} : 0 \leq \xi_0\}.$$

The standard form *primal SOCP problem* is

$$\begin{array}{ll} \text{minimize} & \sum_{j=1}^{\nu} {c^j}^\top x^j \\ \text{subject to} & \sum_{j=1}^{\nu} A^j x^j = b, \\ & x^j \in \mathcal{Q}_{n_j}, \quad j = 1, \ldots, \nu. \end{array} \tag{SOCP-P}$$

Here, the optimization variable is $x = (x^1; \ldots; x^\nu)$, where $x^j \in \mathbb{R}^{n_j}$ for some positive integers n_j, $j = 1, \ldots, \nu$. Moreover, $c^j \in \mathbb{R}^{n_j}$, $A^j \in \mathbb{R}^{m \times n_j}$ and $b \in \mathbb{R}^m$ for some positive integer m. To put this in words, in an SOCP one minimizes a linear function over the intersection of an affine linear space (first constraint) with the Cartesian product of second-order cones (second constraint). Associated with the above primal problem is the *dual problem*

$$\begin{array}{ll} \text{maximize} & b^\top y \\ \text{subject to} & A_j^\top y + z^j = c_j, \quad j = 1, \ldots, \nu, \\ & z^j \in \mathcal{Q}_{n_j}, \quad j = 1, \ldots, \nu \end{array} \tag{SOCP-D}$$

in the optimization variables y and $z = (z^1; \ldots; z^\nu)$. *Primal-dual* interior-point methods solve both the primal and the dual problem at the same time. In the following section we will cast the QCQPs from the last section into the dual SOCP form (SOCP-D).

Primal-dual methods have a pretty nifty feature [2]: For primal feasible x and dual feasible y and z, it always holds true that $c^\top x - b^\top y = z^\top x \geq 0$. Here, $c = (c^1; \ldots; c^\nu)$. Moreover, if the problem is both *strictly primal feasible* and *strictly dual feasible*, i.e., there are primal feasible $x = (x^1; \ldots; x^\nu)$ with $x^j \in \mathcal{Q}^\circ_{n_j}$ (the interior of the cone), $j = 1, \ldots, \nu$, and dual feasible y and $z = (z^1; \ldots; z^\nu)$ with $z^j \in \mathcal{Q}^\circ_{n_j}$, $j = 1, \ldots, \nu$, then there exist optimal solutions x^* of the primal problem and y^* and z^* of the dual problem, and $c^\top x^* - b^\top y^* = {z^*}^\top x^* = 0$. If one inspects the SOCP formulations of (QCQP_p) in the following section, one sees that in these cases $-b^\top y^*$ is actually equal to the minimum of (QCQP_p). Thus, by computing the *duality gap* $c^\top x - b^\top y$ for primal feasible x and dual feasible y, we could in principle get an error estimate for the minimum of (QCQP_p) (and thus (D-OPT_p)).

Reformulating our QCQPs as SOCPs has several advantages. As far as we know, the only publically available software package that can solve QCQPs directly is the commercial solver MOSEK [42]. However, MOSEK's QCQP solver is less efficient than its SOCP solver, and even its authors recommend reformulating QCQPs as SOCPs [43]. On the other hand, there are a number of free software packages available that can solve SOCPs [54, 63]. Further, notice that there occur matrices with tensor-product structure in the constraints of the QCQPs from the previous section. In each of the QCQPs there are at least $|\mathcal{X}^+|$ of these matrices, which have size $N \times N$ and are dense. A software package without special data structures for such matrices would therefore need $O(N^3)$ memory (assuming $|\mathcal{X}^+| = O(N)$). In contrast to this, an inspection of the SOCP formulations below shows that these need only $O(N^2)$ memory.

5.5 SOCP formulation of the discrete problems

5.5.1 General strategy to rewrite QCQPs as SOCPs

Notice that a QCQP can always be rewritten as the optimization of a linear function subject to quadratic constraints: The problem

$$\begin{array}{ll} \text{minimize} & Q(x) \\ \text{subject to} & Q_j(x) \le 0, \quad j = 1, \dots, \nu \end{array}$$

in the optimization variable x with quadratic functions Q and Q_j is equivalent to the problem

$$\begin{array}{ll} \text{minimize} & t \\ \text{subject to} & Q(x) - t \le 0, \\ & Q_j(x) \le 0, \quad j = 1, \dots, \nu \end{array}$$

in the optimization variables t and x. It remains to rewrite the quadratic constraints as second-order cone constraints.

A quadratic constraint of the form

$$x^\top B^\top B x + 2q^\top x + r \le 0$$

with $B \in \mathbb{R}^{m \times N}$, $q \in \mathbb{R}^N$ and $r \in \mathbb{R}$ can be written as a second-order cone constraint as in (SOCP-D) as follows. We have

$$
\begin{aligned}
& x^\top B^\top Bx + 2q^\top x + r \le 0 \\
\Leftrightarrow\quad & \|Bx\|^2 \le -(2q^\top x + r) \\
\Leftrightarrow\quad & \|Bx\|^2 \le \frac{1}{4}\left(\left(1-(2q^\top x + r)\right)^2 - \left(1+(2q^\top x + r)\right)^2\right) \\
\Leftrightarrow\quad & \|Bx\|^2 + \frac{1}{4}\left(1+(2q^\top x + r)\right)^2 \le \frac{1}{4}\left(1-(2q^\top x + r)\right)^2 \\
\Leftrightarrow\quad & \left\| \begin{pmatrix} \frac{1}{2}(1 + 2q^\top x + r) \\ Bx \end{pmatrix} \right\| \le \frac{1}{2}(1 - 2q^\top x - r) \\
\Leftrightarrow\quad & \left\| \begin{pmatrix} q^\top \\ B \end{pmatrix} x + \begin{pmatrix} \frac{1}{2}(1+r) \\ 0_{\mathbb{R}^m} \end{pmatrix} \right\| \le -q^\top x + \frac{1}{2}(1-r).
\end{aligned}
$$

Introducing new variables $\zeta_0 = -q^\top x + \frac{1}{2}(1-r)$ and $\zeta = \begin{pmatrix} q^\top \\ B \end{pmatrix} x + \begin{pmatrix} \frac{1}{2}(1+r) \\ 0_{\mathbb{R}^m} \end{pmatrix}$,

$$
\begin{aligned}
& x^\top B^\top Bx + 2q^\top x + r \le 0 \\
\Leftrightarrow\quad & \begin{cases} \begin{pmatrix} q^\top \\ -q^\top \\ -B \end{pmatrix} x + \begin{pmatrix} \zeta_0 \\ \zeta \end{pmatrix} = \begin{pmatrix} \frac{1}{2}(1-r) \\ \frac{1}{2}(1+r) \\ 0_{\mathbb{R}^m} \end{pmatrix}, \\ \|\zeta\| \le \zeta_0 \end{cases} \\
\Leftrightarrow\quad & \begin{cases} \begin{pmatrix} q^\top \\ -q^\top \\ -B \end{pmatrix} x + \begin{pmatrix} \zeta_0 \\ \zeta \end{pmatrix} = \begin{pmatrix} \frac{1}{2}(1-r) \\ \frac{1}{2}(1+r) \\ 0_{\mathbb{R}^m} \end{pmatrix}, \\ (\zeta_0; \zeta) \in \mathcal{Q}_{m+2}. \end{cases}
\end{aligned}
$$

This is exactly the form of the constraints in (SOCP-D).

A constraint of the form

$$x^\top B^\top Bx + r \le 0$$

can be reformulated in an even easier way:

$$
\begin{aligned}
x^\top B^\top Bx + r \le 0 \quad \Leftrightarrow\quad & \|Bx\| \le (-r)^{1/2} \\
\Leftrightarrow\quad & \begin{cases} \begin{pmatrix} 0_{\mathbb{R}^{1\times N}} \\ -B \end{pmatrix} x + \begin{pmatrix} \zeta_0 \\ \zeta \end{pmatrix} = \begin{pmatrix} (-r)^{1/2} \\ 0 \end{pmatrix}, \\ \|\zeta\| \le \zeta_0 \end{cases} \\
\Leftrightarrow\quad & \begin{cases} \begin{pmatrix} 0_{\mathbb{R}^{1\times N}} \\ -B \end{pmatrix} x + \begin{pmatrix} \zeta_0 \\ \zeta \end{pmatrix} = \begin{pmatrix} (-r)^{1/2} \\ 0 \end{pmatrix}, \\ (\zeta_0; \zeta) \in \mathcal{Q}_{m+1}. \end{cases}
\end{aligned}
$$

5.5.2 $p = 2$, rectangle rule

Let S be a matrix such that

$$S^\top S = \sum_{j \in K \cap \mathcal{X}} (\gamma_j \gamma_j^\top + \sigma_j \sigma_j^\top) h_{\vartheta_j}.$$

Such a matrix S can for example be obtained via eigenvalue decomposition of the right hand side. For example, if $K = \partial\mathbb{D}$ and $\mathcal{X} = \{\mathrm{e}^{i\vartheta_j} : \vartheta_j = j\pi/d,\ j = 1, \ldots, 2d\}$, $d > 1$, then $\sum_{j \in K \cap \mathcal{X}} (\gamma_j \gamma_j^\top + \sigma_j \sigma_j^\top) h_{\vartheta_j} = 2\pi I_N$, where I_N is the identity matrix in $\mathbb{R}^{N \times N}$. In this case, we can take $S = \sqrt{2\pi} I_N$. In the general case, due to the symmetry of K and $\mathcal{X}$,

$$\begin{aligned}\sum_{j \in K \cap \mathcal{X}} (\gamma_j \gamma_j^\top + \sigma_j \sigma_j^\top) h_{\vartheta_j} = 2 &\sum_{j \in K \cap \mathcal{X}^+ \setminus \{\mathrm{e}^{i0}, \mathrm{e}^{i\pi}\}} (\gamma_j \gamma_j^\top + \sigma_j \sigma_j^\top) h_{\vartheta_j} \\ &+ \sum_{j \in K \cap \mathcal{X}^+ \cap \{\mathrm{e}^{i0}, \mathrm{e}^{i\pi}\}} (\gamma_j \gamma_j^\top + \sigma_j \sigma_j^\top) h_{\vartheta_j}.\end{aligned}$$

Since the rank of this matrix is smaller or equal to $2|K \cap \mathcal{X}^+|$, we can find $S \in \mathbb{R}^{m \times N}$ for some $m \leq 2|K \cap \mathcal{X}^+|$. Then ($\mathrm{QCQP}_2$) is equivalent to

$$\begin{aligned}&\text{minimize} && t \\ &\text{subject to} && \alpha^\top S^\top S \alpha + 2q^\top \alpha + \left(\sum_{j \in K \cap \mathcal{X}} |\varphi_j|^2 h_{\vartheta_j} \right) - t \leq 0, \\ &&& \alpha^\top \left(\gamma_j \gamma_j^\top + \sigma_j \sigma_j^\top \right) \alpha - g_j^2 \leq 0, \quad j \in \mathcal{X}^+,\end{aligned}$$

which we again rewrite as

$$\begin{aligned}&\text{maximize} && -t \\ &\text{subject to} && \begin{pmatrix} t \\ \alpha \end{pmatrix}^\top (0_{\mathbb{R}^m}\ \ S)^\top (0_{\mathbb{R}^m}\ \ S) \begin{pmatrix} t \\ \alpha \end{pmatrix} \\ &&& \quad + \begin{pmatrix} -1 \\ 2q \end{pmatrix}^\top \begin{pmatrix} t \\ \alpha \end{pmatrix} + \left(\sum_{j \in K \cap \mathcal{X}} |\varphi_j|^2 h_{\vartheta_j} \right) \leq 0, \\ &&& \begin{pmatrix} t \\ \alpha \end{pmatrix}^\top \begin{pmatrix} 0 & \gamma_j^\top \\ 0 & \sigma_j^\top \end{pmatrix}^\top \begin{pmatrix} 0 & \gamma_j^\top \\ 0 & \sigma_j^\top \end{pmatrix} \begin{pmatrix} t \\ \alpha \end{pmatrix} - g_j^2 \leq 0, \quad j \in \mathcal{X}^+.\end{aligned}$$

From this, the (dual) SOCP formulation is

$$
\begin{aligned}
\text{max.}\quad & \begin{pmatrix} -1 \\ 0_{\mathbb{R}^N} \end{pmatrix}^\top \begin{pmatrix} t \\ \alpha \end{pmatrix} \\
\text{s.t.}\quad & \begin{pmatrix} -\frac{1}{2} & q^\top \\ \frac{1}{2} & -q^\top \\ 0_{\mathbb{R}^m} & -S \end{pmatrix} \begin{pmatrix} t \\ \alpha \end{pmatrix} + z^K = \begin{pmatrix} \frac{1}{2}\left(1 - \sum_{j\in K\cap\mathcal{X}} |\varphi_j|^2 h_{\vartheta_j}\right) \\ \frac{1}{2}\left(1 + \sum_{j\in K\cap\mathcal{X}} |\varphi_j|^2 h_{\vartheta_j}\right) \\ 0_{\mathbb{R}^m} \end{pmatrix}, \\
& z^K \in \mathcal{Q}_{m+2}, \\
& \begin{pmatrix} 0 & 0_{\mathbb{R}^{1\times N}} \\ 0 & -\gamma_j^\top \\ 0 & -\sigma_j^\top \end{pmatrix} \begin{pmatrix} t \\ \alpha \end{pmatrix} + z^j = \begin{pmatrix} g_j \\ 0 \\ 0 \end{pmatrix}, \quad j \in \mathcal{X}^+, \\
& z^j \in \mathcal{Q}_3, \quad j \in \mathcal{X}^+.
\end{aligned}
\tag{SOCP$_2$}
$$

5.5.3 $p = 2$, exact quadrature

Let $\widetilde{S}$ be a matrix such that $\widetilde{S}^\top \widetilde{S} = \widetilde{M}$. We can choose $\widetilde{S} \in \mathbb{R}^{m\times N}$ for some $m \leq N$. As in the last subsection, (QCQP$_{\widetilde{2}}$) can be rewritten as

$$
\begin{aligned}
\text{max.}\quad & \begin{pmatrix} -1 \\ 0_{\mathbb{R}^N} \end{pmatrix}^\top \begin{pmatrix} t \\ \alpha \end{pmatrix} \\
\text{s.t.}\quad & \begin{pmatrix} -\frac{1}{2} & \widetilde{q}^\top \\ \frac{1}{2} & -\widetilde{q}^\top \\ 0_{\mathbb{R}^m} & -\widetilde{S} \end{pmatrix} \begin{pmatrix} t \\ \alpha \end{pmatrix} + z^K = \begin{pmatrix} \frac{1}{2}\left(1 - \left(\sum_{k,l=-N_\varphi}^{N_\varphi-1} m_{k-l}\beta_k\beta_l\right)\right) \\ \frac{1}{2}\left(1 + \left(\sum_{k,l=-N_\varphi}^{N_\varphi-1} m_{k-l}\beta_k\beta_l\right)\right) \\ 0_{\mathbb{R}^m} \end{pmatrix}, \\
& z^K \in \mathcal{Q}_{m+2}, \\
& \begin{pmatrix} 0 & 0_{\mathbb{R}^{1\times N}} \\ 0 & -\gamma_j^\top \\ 0 & -\sigma_j^\top \end{pmatrix} \begin{pmatrix} t \\ \alpha \end{pmatrix} + z^K = \begin{pmatrix} g_j \\ 0 \\ 0 \end{pmatrix}, \quad j \in \mathcal{X}^+, \\
& z^j \in \mathcal{Q}_3, \quad j \in \mathcal{X}^+.
\end{aligned}
\tag{SOCP$_{\widetilde{2}}$}
$$

5.5.4 $p = \infty$

Using

$$
\gamma_j\gamma_j^\top + \sigma_j\sigma_j^\top = \begin{pmatrix} \gamma_j^\top \\ \sigma_j^\top \end{pmatrix}^\top \begin{pmatrix} \gamma_j^\top \\ \sigma_j^\top \end{pmatrix}
$$

and recalling that the optimization variable is $(t;\alpha)$ we rewrite (QCQP$_\infty$) as

$$
\begin{aligned}
\text{maximize}\quad & -t \\
\text{subject to}\quad & \begin{pmatrix} t \\ \alpha \end{pmatrix}^{\top} \begin{pmatrix} 0 & \gamma_j^{\top} \\ 0 & \sigma_j^{\top} \end{pmatrix}^{\top} \begin{pmatrix} 0 & \gamma_j^{\top} \\ 0 & \sigma_j^{\top} \end{pmatrix} \begin{pmatrix} t \\ \alpha \end{pmatrix} \\
& \qquad + \begin{pmatrix} -1 \\ 2q^{(j)} \end{pmatrix}^{\top} \begin{pmatrix} t \\ \alpha \end{pmatrix} + |\varphi_j|^2 \leq 0, \quad j \in K \cap \mathcal{X}^+, \\
& \begin{pmatrix} t \\ \alpha \end{pmatrix}^{\top} \begin{pmatrix} 0 & \gamma_j^{\top} \\ 0 & \sigma_j^{\top} \end{pmatrix}^{\top} \begin{pmatrix} 0 & \gamma_j^{\top} \\ 0 & \sigma_j^{\top} \end{pmatrix} \begin{pmatrix} t \\ \alpha \end{pmatrix} - g_j^2 \leq 0, \quad j \in \mathcal{X}^+.
\end{aligned}
$$

Using the computations from the last subsection we get the (dual) SOCP formulation

$$
\begin{aligned}
\text{max.}\quad & \begin{pmatrix} -1 \\ 0_{\mathbb{R}^N} \end{pmatrix}^{\top} \begin{pmatrix} t \\ \alpha \end{pmatrix} \\
\text{s.t.}\quad & \begin{pmatrix} -\frac{1}{2} & {q^{(j)}}^{\top} \\ \frac{1}{2} & -{q^{(j)}}^{\top} \\ 0 & -\gamma_j^{\top} \\ 0 & -\sigma_j^{\top} \end{pmatrix} \begin{pmatrix} t \\ \alpha \end{pmatrix} + z^K = \begin{pmatrix} \frac{1}{2}(1-|\varphi_j|^2) \\ \frac{1}{2}(1+|\varphi_j|^2) \\ 0 \\ 0 \end{pmatrix}, \quad j \in K \cap \mathcal{X}^+, \\
& z^{K,j} \in \mathcal{Q}_4, \quad j \in K \cap \mathcal{X}^+, \\
& \begin{pmatrix} 0 & 0_{\mathbb{R}^{1\times N}} \\ 0 & -\gamma_j^{\top} \\ 0 & -\sigma_j^{\top} \end{pmatrix} \begin{pmatrix} t \\ \alpha \end{pmatrix} + z^j = \begin{pmatrix} g_j \\ 0 \\ 0 \end{pmatrix}, \quad j \in \mathcal{X}^+, \\
& z^j \in \mathcal{Q}_3, \quad j \in \mathcal{X}^+.
\end{aligned}
\tag{SOCP$_\infty$}
$$

5.6 Numerical experiments

We tried several software packages that can solve SOCPs via interior point methods: SeDuMi [54], SDPT3 [63] and MOSEK [42]. Both SeDuMi and SDPT3 are open source software (GPLv2), while MOSEK is a commercial package. An independent benchmarking of these (and other) packages has been done in [41].

After some tests we decided to use a slightly modified version of SDPT3 for our computations. MOSEK produced rather inaccurate results for larger problem sizes, while both SeDuMi and SDPT3 delivered solutions with good accuracy. Without modifications, SDPT3 was about 1.5 to 3 times slower than SeDuMi for our problems. Unfortunately, all three packages are optimized to perform fast on *sparse* problems. Because our problems are not sparse, all packages produce a significant amount of overhead. Moreover, only MOSEK is parallelized.

The advantage of SDPT3 is that it is mainly written in the MATLAB programming language (with just a few parts in C). By far the most time-consuming step in the algorithms used by SDPT3, SeDuMi and MOSEK is the multiplication of certain large matrices. By simply converting the data type of these matrices from sparse to dense before multiplication we could achieve a speed-up of around 5 to 10 on a single-processor single-core system and around 25 to 50 on a system with 8 cores. The changes we made amounted to less than 10 lines of code and still left the function of SDPT3 completely general. Matrix-matrix multiplication of dense matrices is not only much more efficient, but also the MATLAB implementation is parallelized for dense matrices, but not for sparse matrices. This is why the speed-up is so much larger for multi-processor systems. Some further problem-specific optimizations and the usage of a highly optimized implementation of the basic linear algebra subprograms (Goto BLAS [26]) resulted in an additional speed-up of around 3. Similar modifications should also be possible for SeDuMi, but since large parts of SeDuMi are written in C, this would amount to substantially more work.

The computation times for different sizes of Example 1 from this section can be found in Table 5.1. Even with our optimizations the two most time-consuming lines of code in SDPT3 are two certain matrix multiplications. For example, the total computation time for the largest problem size in Table 5.1 was 578 seconds. About 345 seconds, i.e., around 60 percent of the time, were spent on these two matrix multiplications.

The main limitation to the problem size is memory. Let us exemplify this by (SOCP_∞). The matrices $A_j^\top$ are stored in one large matrix $A^\top = (A_1^\top; \ldots; A_\nu^\top)$. The matrix $A^\top$ has $N+1$ columns, where we recall that N is the dimension of the discrete space $\mathcal{H}_N^\infty(\mathbb{D})$. For every point in $K \cap \mathcal{X}^+$ we get 4 rows and for every point in $\mathcal{X}^+$ we get 3 rows in $A^\top$. Consider Example 1 below with $N = 2^{12} = 4096$ and $d = 2^{12}$ (see (5.13) below), i.e., $|\mathcal{X}^+| = 4097$. In this particular example we have $|K \cap \mathcal{X}^+| = (|\mathcal{X}^+| + 1)/2 = 2049$, so the number of rows in $A^\top$ is

$$4|K \cap \mathcal{X}^+| + 3|\mathcal{X}^+| = 4 \cdot 2049 + 3 \cdot 4097 = 20487.$$

The number of columns is $N + 1 = 4097$. Each element of $A^\top$ needs 8 Bytes, so the total amount of memory needed just to store $A^\top$ is $4 \cdot 4097 \cdot 20487 = 671481912$ Bytes (≈ 0.63 GByte). Unfortunately, SDPT3 is not optimized for memory efficiency and keeps several copies of $A^\top$ in memory. The amount of memory needed for this example in our modified version of SDPT3 (modified 2) is actually around 4.8 GByte on the system that we used. (This is the maximal total memory usage including all temporary results and the memory needed by MATLAB itself.) The unmodified version of SDPT3 requires even more memory, because it uses sparse data structures for matrices that (for our problems) contain only few zero elements. For example, the matrix $A^\top$ from above needs 1073938240 Bytes (≈ 1.00 GByte) when stored in sparse format.

5.6.1 Example 1: Artificial example

As a test problem we took the function φ in the top row of Figure 5.1, $g \equiv 1$ and $K = \{e^{i\vartheta} : \vartheta \in [\frac{\pi}{4}, \frac{3\pi}{4}] \cup [\frac{-3\pi}{4}, \frac{-\pi}{4}]\}$. We solved a series of problems for $p = 2$ with exact quadrature ($N_\varphi = 2^{14}$) and $p = \infty$. We varied the dimension of the space $\mathcal{H}^\infty_N(\mathbb{D})$ by taking $N \in \{2^4, 2^5, \dots, 2^{12}\}$. For each N we solved the problem for several grids of the form

$$\mathcal{X}_d = \left\{ e^{i0\pi/d}, e^{i1\pi/d}, \dots, e^{i(2d-1)\pi/d} \right\}. \tag{5.13}$$

We took $d \in \{2^{\log_2 N}, \dots, 2^{14}\}$.

The minima of the optimization problem with different N and d are shown in Table 5.2. The resulting approximation f^* for $p = 2$ is shown in the middle row of Figure 5.1, and the resulting approximation f^* for $p = \infty$ is shown in the bottom row of Figure 5.1. (In both cases we used $N = 2^{12}$, $d = 2^{14}$). One can see nicely that the solution shows the properties of Theorem 4.4: The absolute value $|f^*|$ is (almost) equal to 1 on $\partial\mathbb{D} \setminus K$. Moreover, in the case $p = \infty$, $|\varphi - f^*|$ is (almost) constant on K.

Let us denote by $\tau^p_{N,d}$ the minimum of the optimization problem with certain p, N and d. Let

$$\delta^p_{N,d} = \left| \tau^p_{N,d} - \tau^p_{N,d/2} \right| \tag{5.14}$$

be the difference between two minima when the number of grid points is doubled. In Figure 5.2 we show how $\delta^p_{N,d}$ behaves for fixed N when we vary

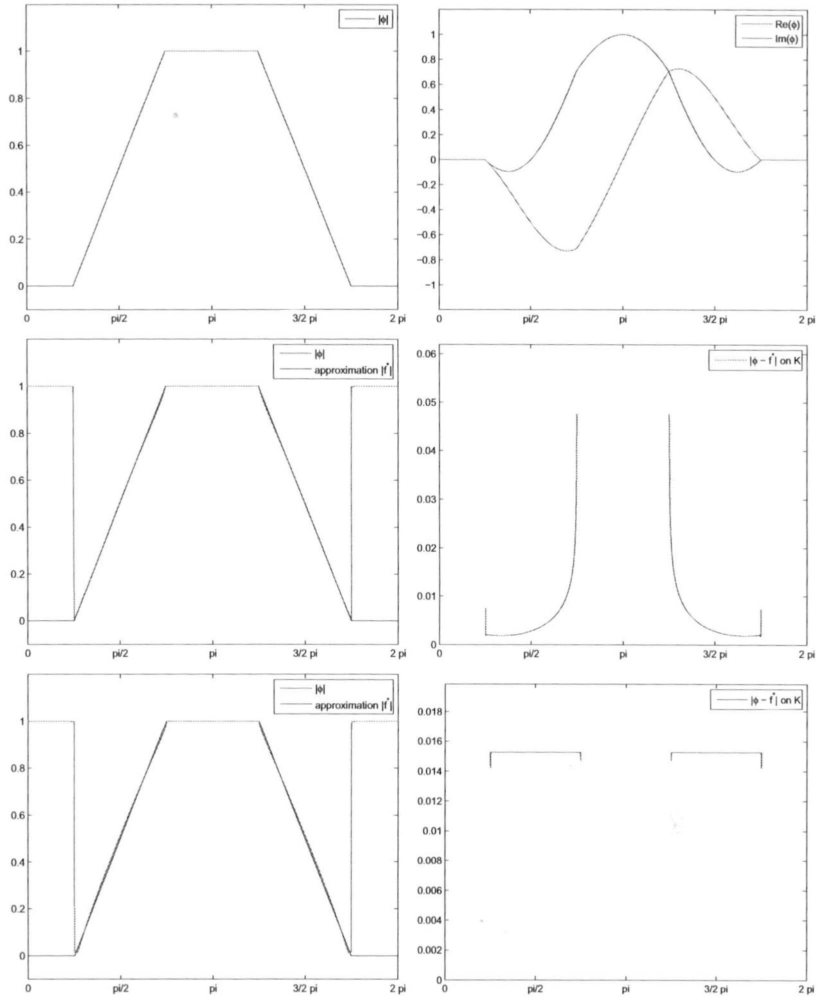

Figure 5.1: Results from Example 1 with $N = 4096$, $d = 16384$. *Top left*: Absolute value of an artificial function φ. *Top right*: Real and imaginary part of φ. *Middle left*: Solution f^* for the case $p = 2$, exact quadrature. *Middle right*: Difference between φ and f^* on K for $p = 2$, exact quadrature. *Bottom left*: Solution f^* for the case $p = \infty$. *Bottom right*: Difference between φ and f^* on K for $p = \infty$.

	$N = 2048$				$N = 4096$	
	$d = 2048$		$d = 4096$		$d = 4096$	
	$p = 2$	$p = \infty$	$p = 2$	$p = \infty$	$p = 2$	$p = \infty$
SeDuMi	3093	6187	6736	24348	24256	51914
SDPT3, original	5578	9357	11913	23544	53755	78986
SDPT3, modified 1	212	320	382	703	1173	1688
SDPT3, modified 1, Goto BLAS	176	257	311	568	959	1397
SDPT3, modified 2, Goto BLAS	73	100	121	205	434	578
#iterations SeDuMi	31	27	37	54	31	29
#iterations SDPT3	28	30	31	37	30	31

Table 5.1: Some computation times (in seconds) for Example 1. Computations were done on a system with two Quad-Core Opteron 2352 processors running at 2.1 GHz, achieving between 25 and 30 Gflops for matrix-matrix multiplication with Goto BLAS. Our MATLAB version was R2007b. For SDPT3 (modified 1) we merely changed the data type of some matrices from sparse to dense before multiplication. SDPT3 (modified 2) contains further optimizations, some of them problem-specific.

d. We observe that both for $p = 2$ and for $p = \infty$, $\delta^p_{N,d}$ behaves like d^{-2}. However, by looking only at the left starting points of the curves in Figure 5.2 (marked by circles) we observe that in the case $p = 2$, also $\delta^p_{d,d}$ behaves like d^{-2}, while in the case $p = \infty$ the decay of $\delta^p_{d,d}$ is significantly slower. This indicates that for $p = 2$ it should suffice to choose $d = N$, while for $p = \infty$ it might be advisable to choose d larger than N.

On the other hand, we consider

$$\Delta^p_d = \left| \tau^p_{d,d} - \tau^p_{d/2,d/2} \right| , \tag{5.15}$$

the difference between two minima when both the number of grid points and the dimension N of the space $\mathcal{H}^\infty_N(\mathbb{D})$ are doubled. Table 5.3 shows the Δ^p_d and the convergence rate $\log_2\left(\Delta^p_d / \Delta^p_{d/2}\right)$, and Figure 5.3 shows a plot. Convergence seems to be approximately linear at least for the cases that we computed.

5.6.2 Example 2: Wideband dispersion compensating mirror

We consider an example from [40]. Here, one is interested in designing a dispersion compensating mirror, i.e., a refractive profile with amplitude reflectivity close to 1 over a large frequency range and a specified phase shift.

(a) $p = 2$, exact quadrature

		d										
		16	32	64	128	256	512	1024	2048	4096	8192	16384
N	16	1.5554e-02	1.8721e-02	1.8997e-02	1.9203e-02	1.9204e-02	1.9206e-02	1.9207e-02	1.9208e-02	1.9208e-02	1.9208e-02	1.9208e-02
	32		1.3531e-02	1.4876e-02	1.4913e-02	1.4955e-02	1.4956e-02	1.4958e-02	1.4959e-02	1.4959e-02	1.4959e-02	1.4959e-02
	64			1.3101e-02	1.3625e-02	1.3640e-02	1.3642e-02	1.3642e-02	1.3643e-02	1.3643e-02	1.3643e-02	1.3643e-02
	128				1.3041e-02	1.3158e-02	1.3178e-02	1.3185e-02	1.3185e-02	1.3185e-02	1.3185e-02	1.3185e-02
	256					1.3032e-02	1.3038e-02	1.3042e-02	1.3043e-02	1.3043e-02	1.3043e-02	1.3043e-02
	512						1.2997e-02	1.3002e-02	1.3003e-02	1.3003e-02	1.3003e-02	1.3003e-02
	1024							1.2988e-02	1.2988e-02	1.2989e-02	1.2989e-02	1.2989e-02
	2048								1.2982e-02	1.2982e-02	1.2982e-02	1.2982e-02
	4096									1.2979e-02	1.2979e-02	1.2979e-02

(b) $p = \infty$

		d										
		16	32	64	128	256	512	1024	2048	4096	8192	16384
N	16	2.1592e-02	2.9678e-02	2.9837e-02	2.9946e-02	3.0049e-02	3.0067e-02	3.0075e-02	3.0075e-02	3.0075e-02	3.0076e-02	3.0076e-02
	32		1.9078e-02	2.2488e-02	2.2844e-02	2.3095e-02	2.3101e-02	2.3103e-02	2.3104e-02	2.3105e-02	2.3105e-02	2.3105e-02
	64			1.7468e-02	1.9056e-02	1.9434e-02	1.9495e-02	1.9517e-02	1.9518e-02	1.9519e-02	1.9519e-02	1.9519e-02
	128				1.6379e-02	1.7273e-02	1.7527e-02	1.7550e-02	1.7551e-02	1.7552e-02	1.7553e-02	1.7553e-02
	256					1.5764e-02	1.6342e-02	1.6454e-02	1.6464e-02	1.6470e-02	1.6470e-02	1.6471e-02
	512						1.5440e-02	1.5815e-02	1.5855e-02	1.5868e-02	1.5869e-02	1.5870e-02
	1024							1.5285e-02	1.5515e-02	1.5536e-02	1.5543e-02	1.5544e-02
	2048								1.5220e-02	1.5354e-02	1.5366e-02	1.5368e-02
	4196									1.5194e-02	1.5269e-02	1.5275e-02

Table 5.2: Minima of the problem from Example 1 with different d and N. (a) Case $p = 2$, exact quadrature. (b) Case $p = \infty$.

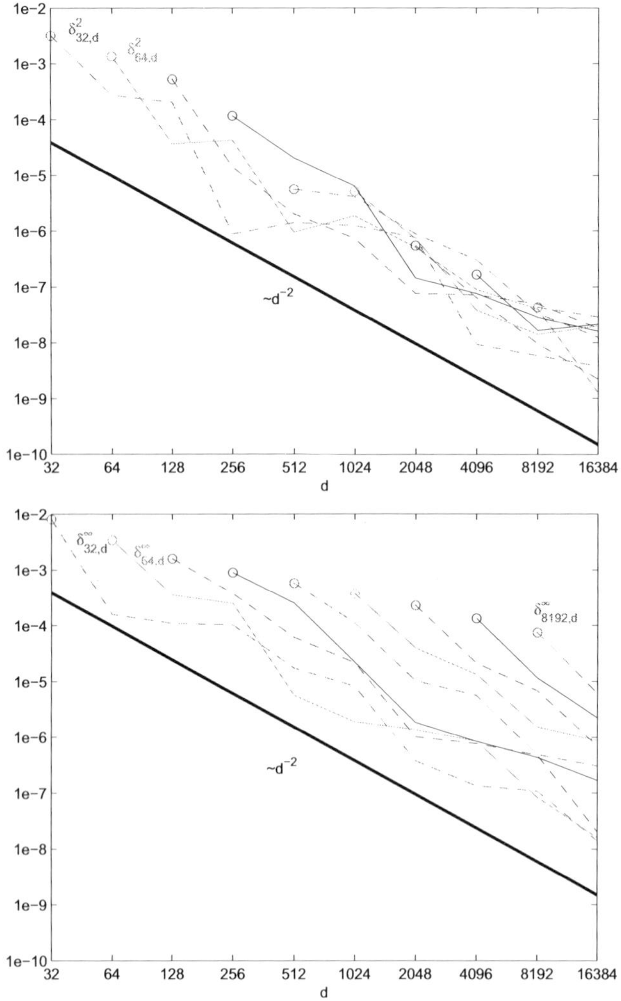

Figure 5.2: *Top*: Case $p = 2$. Each curve shows how $\delta_{N,d}^{2}$ (see equation (5.14)) from Example 1 varies when N is fixed and d is varied. The starting point of each curve is marked by a circle. The curve starting at $d = 32$ is $\delta_{32,d}^{2}$, the curve starting at $d = 64$ is $\delta_{64,d}^{2}$, and so on. The solid black line indicates a decay of d^{-2}. *Bottom*: Case $p = \infty$. Each curve shows how $\delta_{N,d}^{\infty}$ varies when N is fixed and d is varied.

	$p=2$		$p=\infty$	
d	Δ_d^2	rate	Δ_d^∞	rate
32	2.0228e-03		2.5140e-03	
64	4.3028e-04	2.2330	1.6102e-03	0.6428
128	6.0013e-05	2.8419	1.0895e-03	0.5636
256	8.6874e-06	2.7883	6.1446e-04	0.8263
512	3.5529e-05	-2.0320	3.2421e-04	0.9224
1024	9.0181e-06	1.9781	1.5537e-04	1.0613
2048	5.7313e-06	0.6540	6.4489e-05	1.2685
4096	2.7908e-06	1.0382	2.5656e-05	1.3298

Table 5.3: Difference between two minima of Example 1 when both the number of grid points and the dimension of the space $\mathcal{H}_N^\infty(\mathbb{D})$ are doubled, see (5.15).

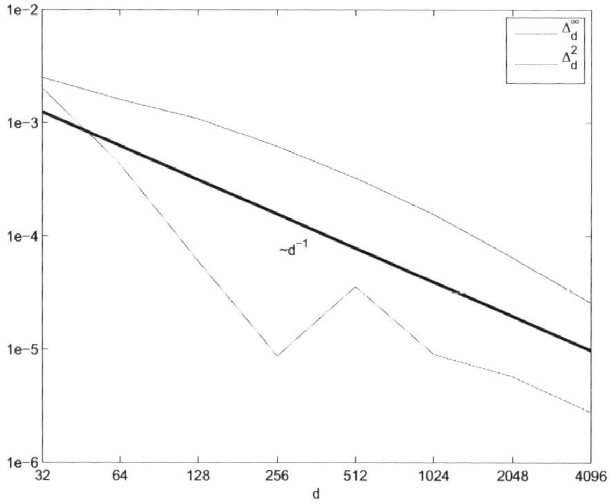

Figure 5.3: The data from Table 5.3. The solid black line indicates a decay of d^{-1}.

The desired phase shift is given as a polynomial around some center frequency ω_0,

$$\phi(\omega) = \sum_{\nu=0}^{k} \frac{1}{\nu!} D_\nu (\omega - \omega_0)^\nu,$$

see Chapter 1. Here, ω is the angular frequency. The relationship between the angular frequency ω and the wavelength λ is given by $\omega = 2\pi c_0/\lambda$, where $c_0 = 299792458\ \mathrm{ms}^{-1}$ is the speed of light in vacuum. The numbers D_ν are the *dispersion coefficients*. We explained in Chapter 1 that in practice, one is only interested in the group delay dispersion

$$GDD(\omega) = \frac{\partial^2}{\partial \omega^2}\phi(\omega) = \sum_{\nu=2}^{k} \frac{1}{(\nu-2)!} D_\nu (\omega - \omega_0)^{\nu-2}.$$

Hence, only the dispersion coefficients for $\nu \geq 2$ are relevant, and there is some freedom in the choice of D_0 and D_1. We consider the reflection coefficient with respect to $\omega = kc_0$, which has the same qualitative properties as the reflection coefficient with respect to k.

We take dispersion coefficients and center frequency from [40, Table 4.1], $D_2 = -50.0\,\mathrm{fs}^2$, $D_3 = 32.2\,\mathrm{fs}^3$, $D_4 = 268.2\,\mathrm{fs}^4$, $D_5 = -62.3\,\mathrm{fs}^5$ and center wavelength $\lambda_0 = 760\,\mathrm{nm}$, that is, $\omega_0 = 2.4785\,\mathrm{fs}^{-1}$. We chose $D_1 = 27\,\mathrm{fs}$ and $D_0 = 0$. The desired reflection coefficient is then

$$R_{\mathrm{desired}}(\omega) = \exp\left(i \sum_{\nu=0}^{5} \frac{1}{\nu!} D_\nu (\omega - \omega_0)^\nu \right),$$

see Figure 5.4. As the interval of interest we take $[550\,\mathrm{nm}, 1300\,\mathrm{nm}]$, or $[1.4490\,\mathrm{fs}^{-1}, 3.4248\,\mathrm{fs}^{-1}]$. In order to transport functions from the half-plane to the disk and back, we use the isometry T_∞ from Theorem 2.12. Together with real symmetry we get

$$K = \{e^{i\vartheta} : \vartheta \in [-2.5734, -1.9334] \cup [1.9334, 2.5734]\}.$$

The resulting function φ is shown in Figure 5.5.

Computations were done with $N = 4096$ and the grid $\mathcal{X}_d$ as in (5.13) with $d = 16384$. In order to use exact quadrature in the case $p = 2$, we need to continue φ from K to the whole circle. We did this in such a way that the resulting φ is smooth on the whole circle.

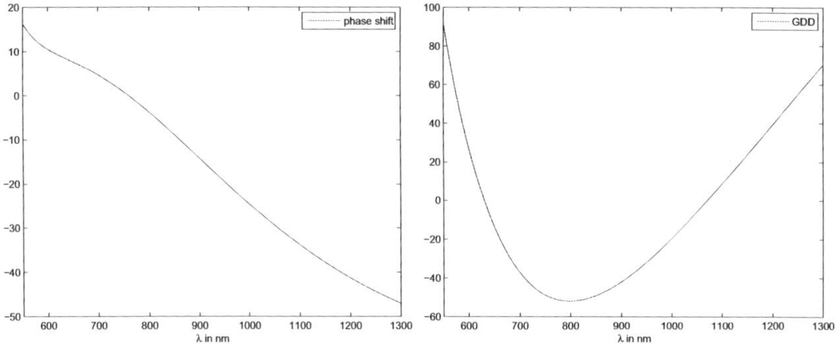

Figure 5.4: Example 2. *Left*: Desired phase shift ϕ. *Right*: Desired group delay dispersion (GDD).

The results are shown in Figure 5.6. Interestingly, the solutions seem to show the property from Theorem 4.4 only on the set K. In the case $p = 2$ we obtained 1.8782e-04 for the minimum of the discrete problem. In the case $p = \infty$ we obtained 2.5771e-04 for the minimum of the discrete problem.

Notice further that on can see the implications of Remark 4.10 in Figure 5.6. We were able to show that for $p = \infty$ the solution f^* is smooth on the sets Γ_1° and Γ_2°, where $\Gamma_1 = \{e^{i\vartheta} \in \partial\mathbb{D} : |f^*(e^{i\vartheta}) - \varphi(e^{i\vartheta})| = \tau^*\}$ and $\Gamma_2 = \{e^{i\vartheta} \in \partial\mathbb{D} : |f^*(e^{i\vartheta})| = g(e^{i\vartheta})\}$. The plot of the GDD indicates that f^* is indeed not smooth on the boundary of Γ_1 and Γ_2. In Example 3, this is even more noticeable.

5.6.3 Example 3: DCM with pump window

We consider an example from the diploma thesis of Felix Grawert [27]. Grawert investigated the design of dispersion compensation mirrors with very challenging restrictions. An example where he failed to obtain a mirror that meets the design goal is the following [27, Section 6.2.1]. He wanted to have a reflectivity larger than 0.999 between 730 nm and 850 nm. Further, he needed a pump window from 672 nm to 682 nm where the reflectivity is smaller than 0.05. Unfortunately, he does not state the desired dispersion coefficients in the high reflectance region. From his graphs we guess $D_3 = -35\,\mathrm{fs}^3$ and $D_2 = -57\,\mathrm{fs}^2$ at a center wavelength of 800 nm. Moreover, we

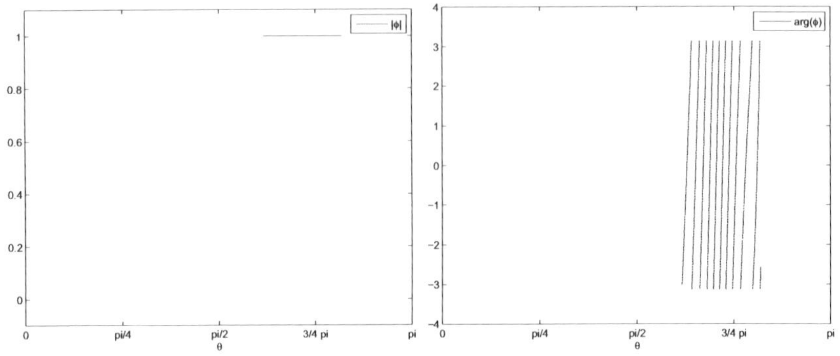

Figure 5.5: Example 2. Desired reflection coefficient transported to the circle in the interval of interest. Due to symmetry we only show the interval $[0, \pi]$. *Left*: absolute value $|\varphi|$. *Right*: argument $\arg(\varphi)$.

chose $D_1 = -27\,\text{fs}$ and $D_0 = 0$.

We modelled this situation as follows. As the desired reflection coefficient we took $r(\omega) = e^{i\sum_{\nu=0}^{3}\frac{1}{\nu!}D_\nu(\omega-\omega_0)^\nu}$ and transported this to the circle as in Example 2 to obtain φ, see Figure 5.7. The set K is the HR region $[730\,\text{nm}, 850\,\text{nm}]$ transported to the circle. In order to take care of the pump window, we transported the interval $[672\,\text{nm}, 682\,\text{nm}]$ to the circle and chose $g = 0.05$ on this interval and $g = 1$ on the rest of the circle. Computations were done for $p = \infty$ with $N = 8192$. For the grid we took the points from $\mathcal{X}_d$, see (5.13), with $d = 8192$, but we added some additional points in and around the HR region and the pump region so that the final grid contained 14517 points.

The result is shown in Figure 5.8. For the minimum we obtained 8.1547e-04, that is, a reflectivity of at least 0.99918 over the HR region, so our solution barely satisfies GRAWERT's requirements. However, since the space of realizable reflection coefficients is quite a bit smaller than $H^\infty(\mathbb{C}^+)$, this is a strong sign that it is not possible at all to design a mirror that meets the design goals.

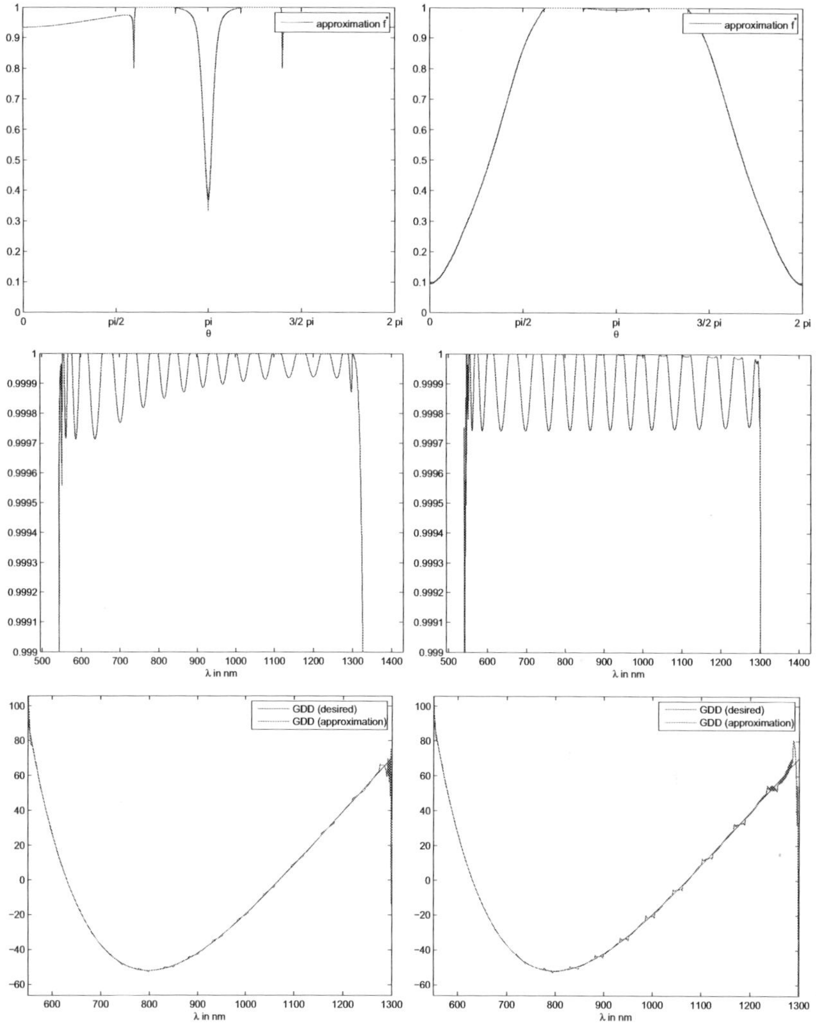

Figure 5.6: Example 2, solution. *Left column*: Case $p = 2$. *Right column*: Case $p = \infty$. *Top row*: Solution f^*. *Middle row*: Solution transported to the real line. *Bottom row*: Desired GDD and GDD of the solution.

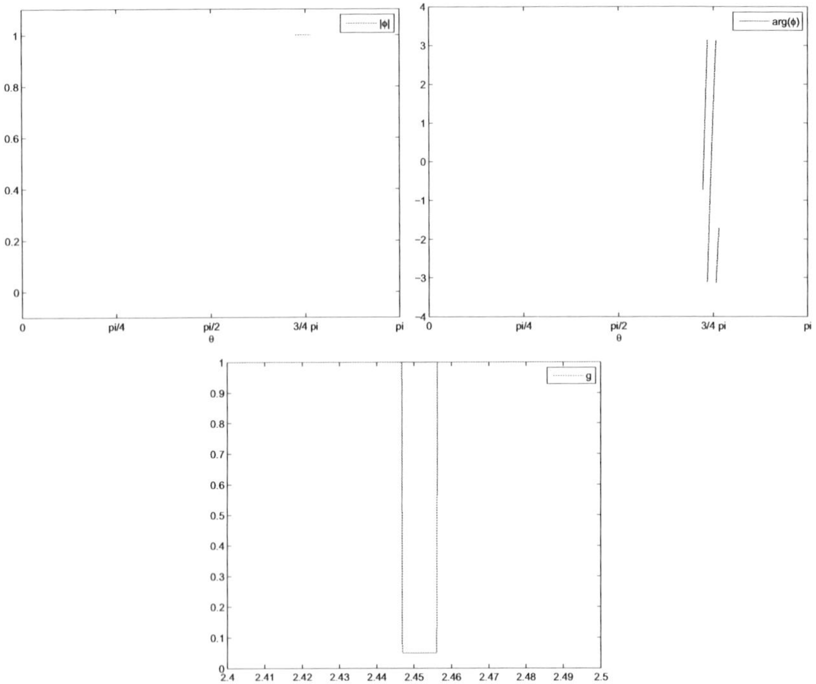

Figure 5.7: Example 3. *Top row*: Desired reflection coefficient transported to the circle, absolute value and argument. *Bottom*: Barrier function g. Since the interval where g is not equal to 1 is so small, only a small subset of the circle is shown.

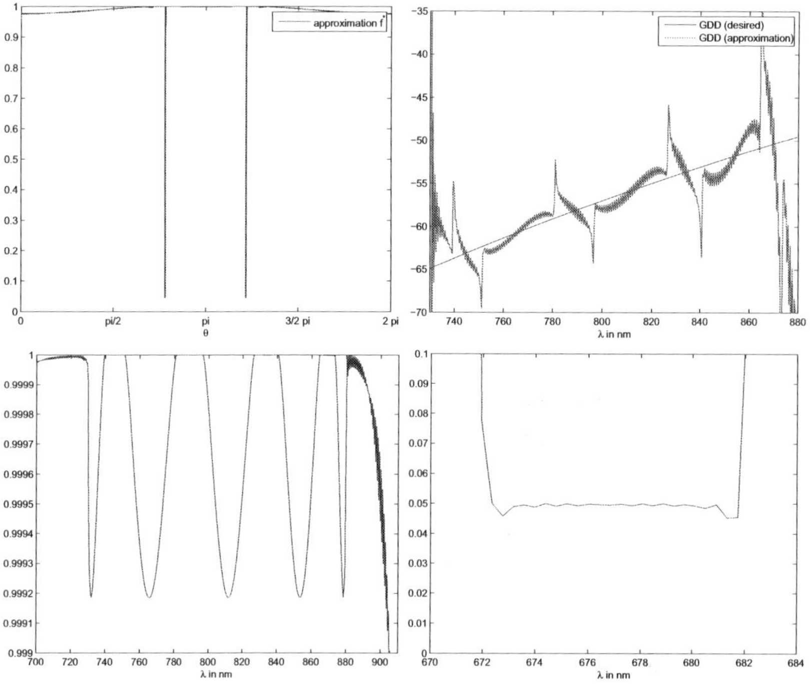

Figure 5.8: Example 3. *Top left*: Solution f^*, absolute value. *Bottom left*: Solution transported to the real line (HR region), absolute value. *Bottom right*: Solution transported to the real line (pump region), absolute value. *Top right*: Desired GDD and GDD of the solution.

Bibliography

[1] Z. S. Agranovich and V. A. Marchenko, *The inverse problem of scattering theory*, Gordon and Breach, 1963.

[2] F. Alizadeh and D. Goldfarb, *Second-order cone programming*, Math. Program., Ser. B, 95 (2003), pp. 3–51.

[3] H. W. Alt, *Lineare Funktionalanalysis*, Springer, 4th ed., 2002.

[4] L. Baratchart and J. Leblond, *Hardy approximation to L^p functions on subsets of the circle with $1 \leq p < \infty$*, Constr. Approx., 14 (1998), pp. 41–56.

[5] L. Baratchart, J. Leblond, and J. R. Partington, *Hardy approximation to L^∞ functions on subsets of the circle*, Constr. Approx., 12 (1996), pp. 423–435.

[6] L. Baratchart, J. Leblond, J. R. Partington, and N. Torkhani, *Robust identification from band-limited data*, IEEE Trans. Automat. Contr., 42 (1997), pp. 1318–1325.

[7] E. J. Beltrami and M. R. Wohlers, *Distributions and the Boundary Values of Analytic Functions*, Academic Press, 1966.

[8] F. A. Berezin and M. A. Shubin, *The Schrödinger equation*, Kluwer, 1991.

[9] M. Born and E. Wolf, *Principles of Optics*, Cambridge University Press, 7th ed., 1999.

[10] P. Brémaud, *Mathematical principles of signal processing*, Springer, 2002.

[11] J. Brunk, *Optimale Beschichtung von Laserspiegeln zur Erzeugung ultrakurzer Laserpulse*, PhD thesis, Universität Karlsruhe, 2006.

[12] R. D. CARMICHAEL AND D. MITROVIĆ, *Distributions and analytic functions*, Longman Scientific and Technical, 1989.

[13] K. CHADAN, D. COLTON, L. PÄIVÄRINTA, AND W. RUNDELL, *An introduction to inverse scattering and inverse spectral problems*, SIAM, 1997.

[14] K. CHADAN AND P. C. SABATIER, *Inverse problems in quantum scattering theory*, Springer, 2nd rev. and enl. ed., 1989.

[15] C. P. CHANG, Y. H. LEE, AND S. Y. WU, *Optimization of a thin-film multilayer design by use of the generalized simulated-annealing method*, Opt. Lett., 15 (1990), pp. 595–597.

[16] E. M. CHIRKA, *Regularity of the boundaries of analytic sets*, Math. USSR Sb., 45 (1983), pp. 291–335.

[17] S. J. COX AND J. R. MCLAUGHLIN, *Extremal eigenvalue problems for composite membranes, I*, Appl. Math. Optim., 22 (1990), pp. 153–167.

[18] P. DEIFT AND E. TRUBOWITZ, *Inverse scattering on the line*, Comm. Pure Appl. Math., 32 (1979), pp. 121–251.

[19] D. C. DOBSON, *Optimal design of periodic antireflective structures for the Helmholtz equation*, Eur. J. Appl. Math., 4 (1993), pp. 321–340.

[20] L. C. EVANS, *Partial Differential Equations*, American Mathematical Society, 1998.

[21] L. D. FADDEEV AND B. SECKLER, *The inverse problem in the quantum theory of scattering*, J. Math. Phys., 4 (1963), pp. 72–104.

[22] R. FLETCHER, *Practical methods of optimization*, Wiley, 1987.

[23] G. B. FOLLAND AND A. SITARAM, *The uncertainty principle: A mathematical survey*, J. Fourier Anal. Appl., 3 (1997), pp. 207–238.

[24] S. A. FURMAN AND A. V. TIKHONRAVOV, *Basics of optics of multilayer systems*, Editions Frontières, 1992.

[25] J. B. GARNETT, *Bounded analytic functions*, Academic Press, 1981.

[26] K. Goto and R. van de Geijn, *High-performance implementation of the level-3 BLAS*, ACM Trans. Math. Softw., 35 (2008), pp. 1–14.

[27] F. Grawert, Diploma thesis, Universität Karlsruhe (TH), 2001.

[28] N. I. Grinberg, *The one-dimensional inverse scattering problem for the wave equation*, Math. USSR Sb., 70 (1991), pp. 557–572.

[29] J. W. Helton and R. E. Howe, *A bang-bang theorem for optimization over spaces of analytic functions*, J. Approx. Theory, 47 (1986), pp. 101–121.

[30] J. W. Helton and D. E. Marshall, *Frequency domain analysis and analytic selections*, Indiana Univ. Math. J., 39 (1990), pp. 157–184.

[31] J. W. Helton and O. Merino, *Classical control using H^∞ methods: theory, optimization, and design*, SIAM, 1998.

[32] K. Hoffman, *Banach spaces of analytic functions*, Dover, 1988.

[33] S. Hui, *Qualitative properties of solutions to H^∞-optimization problems*, J. Func. Anal., 75 (1987), pp. 323–348.

[34] F. Jarre, M. Kocvara, and J. Zowe, *Optimal truss design by interior-point methods*, SIAM J. Optim., 8 (1998), pp. 1084–1107.

[35] A. P. Joglekar, H. Liu, E. Meyhöfer, G. Mourou, and A. J. Hunt, *Optics at critical intensity: Applications to nanomorphing*, Proc. Natl. Acad. Sci. USA, 101 (2004), pp. 5856–5861.

[36] Y. Katznelson, *An Introduction to Harmonic Analysis*, Wiley, 1968.

[37] H. N. Kritikos, D. L. Jaggard, and D. B. Ge, *Numeric reconstruction of smooth dielectric profiles*, Proc. IEEE, 70 (1982), pp. 295–297.

[38] H. A. Macleod, *Thin-film optical filters*, Adam Hilger Ltd, London, 1969.

[39] S. Martin, J. Rivory, and M. Schoenauer, *Synthesis of optical multilayer systems using genetic algorithms*, Appl. Opt., 34 (1995), pp. 2247–2254.

[40] N. Matuschek, *Theory and Design of Double-Chirped Mirrors*, PhD thesis, ETH Zürich, 1999.

[41] H. D. Mittelmann, *An independent benchmarking of SDP and SOCP solvers*, Math. Program., Ser. B, 95 (2003), pp. 407–430.

[42] MOSEK ApS, *The MOSEK optimization software.* `http://www.mosek.com`.

[43] ——, *The MOSEK optimization tools manual, Version 5*, 2008.

[44] Z. Nehari, *Conformal mapping*, McGraw-Hill, 1952.

[45] Y. Nesterov and A. Nemirovskii, *Interior-Point Polynomial Algorithms in Convex Programming*, SIAM, 1987.

[46] J. N. Pandey, *The Hilbert Transform of Schwarz Distributions and Applications*, Wiley, 1996.

[47] C. Papachristos and P. Frangos, *Design of corrugated optical waveguide filters through a direct numerical solution of the coupled Gel'fand–Levitan–Marchenko integral equations*, J. Opt. Soc. Am. A, 19 (2002), pp. 1005–1012.

[48] V. Pervak, A. V. Tikhonravov, M. K. Trubetskov, S. Naumov, F. Krausz, and A. Apolonski, *1.5-octave chirped mirror for pulse compression down to sub-3 fs*, Appl. Phys. B, 87 (2007), pp. 5–12.

[49] W. Rudin, *Real and complex analysis*, McGraw-Hill, 3rd ed., 1987.

[50] C. Rullière, *Femtosecond Laser Pulses*, Springer, 2nd ed., 2005.

[51] T. R. Schibli, O. Kuzucu, J.-W. Kim, E. P. Ippen, J. G. Fujimoto, F. X. Kaertner, V. Scheuer, and G. Angelow, *Toward single-cycle laser systems*, IEEE Sel. Top. Quantum Electron., 9 (2003), pp. 990–1001.

[52] R. Scipőcs, K. Ferencz, C. Spielmann, and F. Krausz, *Chirped multilayer coatings for broadband dispersion control in femtosecond lasers*, Opt. Lett., 19 (1994), pp. 201–203.

[53] R. Scipőcs and A. Kőházi-Kis, *Theory and design of chirped dielectric laser mirrors*, Appl. Phys. B, 65 (1997), pp. 115–135.

[54] J. F. STURM, *Using SeDuMi 1.02, A Matlab toolbox for optimization over symmetric cones*, Optimization Methods and Software, 11 (1999), pp. 625–653.

[55] J. F. STURM, *Implementation of interior point methods for mixed semidefinite and second order cone optimization problems*, Optimization Methods and Software, 17 (2002), pp. 1105–1154.

[56] J. SYLVESTER, *Layer stripping*, in Surveys on Solution Methods for Inverse Problems, D. Colton, H. Engl, A. K. Louis, J. R. McLaughlin, and W. Rundell, eds., Springer, 2000, pp. 83–106. Available online at `http://www.math.washington.edu/~sylvest/papers/layerstrip.pdf`.

[57] J. SYLVESTER AND D. WINEBRENNER, *Linear and nonlinear inverse scattering*, SIAM J. Appl. Math., 59 (1998), pp. 669–699.

[58] J. SYLVESTER, D. WINEBRENNER, AND F. GYLYS-COLWELL, *Layer stripping for the Helmholtz equation*, SIAM J. Appl. Math., 56 (1996), pp. 736–754.

[59] G. J. TEARNEY, M. E. BREZINSKI, B. E. BOUMA, S. A. BOPPART, C. PITRIS, J. F. SOUTHERN, AND J. G. FUJIMOTO, *In vivo endoscopic optical biopsy with optical coherence tomography*, Science, 276 (1997), pp. 2037–2039.

[60] A. THELEN, *Design of optical interference coatings*, McGraw-Hill, 1988.

[61] A. V. TIKHONRAVOV AND M. K. TRUBETSKOV, *OptiLayer Thin Film Software*. `http://www.optilayer.com`.

[62] A. V. TIKHONRAVOV, M. K. TRUBETSKOV, AND G. W. DEBELL, *Application of the needle optimization technique to the design of optical coatings*, Appl. Opt., 35 (1996), pp. 5493–5508.

[63] K. C. TOH, R. H. TÜTÜNCÜ, AND M. J. TODD, *On the implementation and usage of SDPT3 – a Matlab software package for semidefinite-quadratic-linear programming, version 4.0*, July 2006. Available online at `http://www.math.nus.edu.sg/~mattohkc/sdpt3.html`.

[64] A. VISINTIN, *Strong convergence results related to strict convexity*, Comm. Part. Diff. Eq., 9 (1984), pp. 439–466.

[65] W. WALTER, *Gewöhnliche Differentialgleichungen*, Springer, 7th ed., 2000.

[66] N. YOUNG, *An introduction to Hilbert space*, Cambridge University Press, 1988.

[67] A. H. ZEWAIL, *Femtochemistry: Atomic-scale dynamics of the chemical bond*, J. Phys. Chem. A, 104 (2000), pp. 5660–5694.

Printed in Dunstable, United Kingdom

91343482R00085